Imagine Being Connected

Stefano Manfroi

IMAGINE BEING CONNECTED

Website: stefano.manfroi.it
Email: stefano@manfroi.it

Publisher Information

America Publishers
Email: info@americapublishers.com
Phone: +1 (617) 334-5774

ISBN Information

Paperback: 978-1-966198-62-8

Cover Design by: Stefano Manfroi

Printed in the United States of America

1st Edition: February, 2025

Author's Biography

My name is Stefano Manfroi, and I am a technician, an IT professional with over forty years of experience. My passion for technology and innovation has always pushed me to stay ahead of the present reality, anticipating changes and pioneering new paths in the world of computer science and telecommunications.

I began my career in 1983 when I purchased the legendary Commodore 64, the first home computer that marked the beginning of my journey into the digital world. From that moment on, I followed and contributed to technological evolution, working with all types of computers and developing skills ranging from programming to managing telecommunication networks.

In 1994, I took a step that would change the way people in my region interacted with technology: I became the first Internet provider in Umbria, offering my fellow citizens in Terni the possibility of connecting to the internet from home for the cost of a local phone call. At a time when the internet in Italy was still an unexplored territory, I had the vision to bring this innovation directly into people's homes. While many were still unaware of the potential of the web, I was already working to make it accessible to everyone.

Being a pioneer doesn't just mean adopting new technologies but foreseeing their impact and knowing how to integrate them into people's daily lives.

I have always believed that technology is not an end in itself but a tool to improve knowledge, communication, and human awareness.

This belief has led me over the years to explore new frontiers between science and the mind, delving into the study of brain waves and the interaction between imagination and technology. If technology can connect the world, can it also connect people with themselves? This question has guided my research into a still little-explored area: the power of imagination and its effect on the human mind.

With ***Imagine Being Connected***, I wanted to combine my experience in computer science and telecommunications with research on the mind's capabilities, creating a bridge between technology, science, and emotion. This book represents a meeting point between my past as a digital pioneer and my ongoing quest to understand human consciousness.

Preface

My journey with *Imagine Being Connected* began with a deep reflection: to investigate, using appropriate means, what happens in the mind when we imagine being in love. This book was not born by chance but is the result of years of study, experience, and research.

It all started with my previous novel, *Imagine Being in Love*. Writing that book was a fascinating journey, where I explored the idea that love could exist solely in the mind, without the need for a physical presence. The concept intrigued me so much that I felt the need to delve deeper, pushing myself to investigate what really happens in the brain when experiencing an imagined emotion.

Thus, a new challenge was born: not just telling a story but also developing a research protocol to scientifically explore imagined love. I began testing and verifying, using **eMind EEG**, an advanced device for monitoring brain waves. My goal was to understand how the brain reacts not only to real experiences but also to those created exclusively in the mind.

The data collected led me to a fascinating discovery: brain waves respond in a measurable way even to emotions and situations created solely by the imagination.

This research gave birth to *Imagine Being Connected*, a novel that is not just a story but a journey through science, technology, and emotion, accompanied by a manual with a true research protocol.

I do not claim to teach anything or provide definitive answers. I simply want to share an experience that has united my technical world with my emotional one, creating a narrative that I hope will inspire anyone who reads it.

Perhaps this book will not change your life, but I hope it will push you to explore your potential. **Because imagining being in love does not require another person: it only requires the ability to dream and let yourself be carried away.**

Where It All Begins

"Two distant lives, two rational yet sensitive minds. Paolo and Francesca do not know each other, yet they share something profound: the search for a greater meaning, the curiosity to explore the boundaries of their minds.

When a simple message introduces them to a technology capable of monitoring thought and imagination, their world begins to change.

What seemed like just a scientific innovation turns into an inner journey, an exploration of emotions and the connection between reality and imagination.

Through **eMind EEG**, Paolo and Francesca will discover that sometimes, you don't need to live an experience to feel it as real. And that, perhaps, what we imagine has more power than we believe..."

Contents

Chapter 1:

Paolo - The Beginning of an Inner Journey

✦ ✦ ✦

Paolo was not the kind of person to get caught up in abstract concepts. To him, reality was made up of numbers, well-defined projects, and problems to be solved. Everything had to be precise, measurable, and clearly defined. Seated in his office, surrounded by stacks of papers and monitors glowing with complex graphs and colorful tables, he felt safe in that rational world he had so carefully built. Every click of the mouse was a small step toward a solution, every data point a confirmation of his competence.

It had been a long day. The project he was working on required precision and concentration, and even the smallest mistake could cost him hours of work and the respect of his superiors. But just as he was about to shut down his computer, exhausted, a notification flashed on the screen, capturing his attention.

A message from Marco, his colleague, always ready to share the latest technological novelties.

"You need to see this, Paolo. It might interest you."

The text was accompanied by a link to a video. Paolo, accustomed to thinking with logic and precision, couldn't help but notice how vague yet intriguing that message was. With a sigh, he clicked on the link, more out of courtesy than genuine curiosity.

He never could have imagined that such a simple action would change his perception of reality.

The video showed a woman, calm and focused, wearing a device on her head—a lightweight band, elegantly and discreetly designed, with a small sensor positioned on her forehead. The narrator's voice, calm and reassuring, explained how the device could detect brain activity and translate it into visible data, showing in real-time how the mind reacted to thoughts and emotions.

Paolo watched skeptically. "Interesting, but what does this have to do with me?" he thought, almost amused by his own reaction.

The video continued, explaining how the device could be used to explore emotional and imaginative states, helping individuals better understand themselves through technology.

"Imagine situations and monitor your brain while doing it."

"Understand yourself better through technology."

That phrase struck him deeply.

The idea of monitoring his own brain activity while imagining scenarios disturbed and fascinated him at the same time. Paolo didn't like to admit it, but lately, he had felt trapped in a cold and predictable routine. He was a successful man, with a solid and rewarding career, but that lingering sense of emptiness that gripped him every evening when he returned home made him question everything.

His life had become a series of mechanical actions, devoid of the spontaneity he had always dreamed of. And yet, he couldn't shake that feeling of detachment, as if he were living in a parallel reality, far from genuine emotions.

Maybe, he thought, there was something in that proposal worth exploring.

That evening, instead of unwinding in front of a TV series, Paolo spent hours reading articles and reviews about **eMind EEG**.

The more he delved into it, the more his curiosity grew.

The idea of monitoring his own mind, analyzing his emotional and mental state, seemed both bizarre and fascinating to him.

"This isn't my thing," he kept repeating to himself. And yet, he couldn't ignore that strange sense of attraction.

There was something about this technology that, though far from his rational and analytical nature, was calling him toward a new discovery.

Before going to bed, he had already placed his order.

He didn't know it yet, but that impulsive action would set in motion a journey that would shake all his certainties.

A journey that would lead him to explore a part of himself he had always ignored.

A journey that, perhaps, would help him truly understand what it meant to be himself.

Chapter 2:

Francesca - The Awakening of Curiosity

✦ ✦ ✦

Francesca, seated in her room bathed in the soft morning light, let herself be enveloped by the silence. She had never been particularly drawn to technological novelties. To her, the world was made of sensations, written words, and deep reflections. Technology was something distant, a necessary background but never essential.

And yet, behind her apparent calm, there was a storm of emotions, a thirst for understanding that rarely found answers in the world around her.

Her apartment, small but cozy, reflected her essence: few objects, but each one rich in meaning. A painting of the sea, an open book of poetry on her desk, a steaming cup of tea near the window, from which she could observe the tranquil garden below.

Her life, though busy, had always followed a natural rhythm, never forcing her beyond her limits.

But that morning, something was different.

It was an indefinable feeling.

It wasn't tied to a particular event, but rather to a door that had always remained closed and now seemed slightly ajar. A subtle yet persistent curiosity.

Almost without realizing it, she opened her computer—an action she usually avoided, preferring to immerse herself in the reassuring scent of paper and ink.

A notification appeared on the screen.

A message from Emma, a friend who loved sharing the latest tech trends.

"Francesca, you need to watch this video. You might find it interesting."

Attached to the message was a link leading to a video about **eMind EEG**.

Francesca had never had a particular interest in technology, but that morning, curiosity got the better of her.

The video showed a serene and focused woman wearing an elegant, lightweight band with a small sensor positioned on her forehead.

The narrator explained how this device could detect brain activity and translate it into visible data to better understand mental and emotional states.

"Imagining and monitoring your brain while doing it is a way to understand yourself better."

Those words struck a chord deep within her.

Francesca felt transported elsewhere.

She had never been someone who sought answers in technology, yet the concept of exploring her own mind through a scientific tool seemed to offer a new perspective on herself.

Maybe, she thought, this device could enrich her inner life.

Maybe she could use it to explore those parts of herself that often slipped away.

To enhance her awareness during meditation. To expand her field of introspection.

As the video continued, her fingers moved automatically across the keyboard. Almost without thinking, she found herself placing an order for the **eMind EEG**.

She wasn't entirely sure why she had made that decision, but deep down, she felt she had to.

It wasn't just an impulse. It was something deeper.

It was a call she could no longer ignore.

That evening, as the sky turned golden and her home wrapped itself in silence, Francesca felt a strange sense of peace mixed with excitement.

She didn't yet know where this journey would take her, but she knew it was the beginning of a new chapter in her life.

A journey that, perhaps, would lead her to a deeper understanding of herself.

Chapter 3:

Paolo - The First Contact

✦ ✦ ✦

That morning, the package had arrived, carefully placed on his doorstep. Paolo picked it up with one hand, surprised by its light weight, and carried it straight to the kitchen. He wasn't usually the type to get excited about a purchase, but there was something different about this one—something that made him eager to open it and see what was inside. Maybe it was curiosity to determine if all that technology really made sense, as promised in the videos he had watched over the past few days. His analytical mind, which usually wasn't easily influenced, now found itself intrigued by the idea of exploring new territory—the realm of the mind and emotions, which he had always considered more elusive and less measurable than numbers and graphs.

He placed the box on the wooden table, his heart pounding with that almost primitive sensation that accompanies the moment of opening something new. With precise movements, typical of his meticulous nature, he began removing the adhesive tape and opening the package. Inside, the device appeared sleek and minimalist: a lightweight headband with a sensor that looked like something out of a science-fiction movie yet maintained a disarmingly simple design. There were no complicated wires or bulky equipment—only the essentials, as if the entire device had been designed to be as inconspicuous as possible. Next to it, the instruction manual promised simple and immediate use, and Paolo, with his typically pragmatic approach, decided to try it out right away without unnecessary hesitation.

After downloading the app onto his tablet, Paolo put on the device. The frontal sensor gently adhered to his skin, almost like a soft embrace, and a small blue light turned on, signaling that the device was ready to be used. That tiny blue light felt almost like a sign—a confirmation that the world of rationality and the world of intuition were about to intersect. Paolo followed the instructions, configuring the initial settings, trying not to let himself be overwhelmed by the slight sense of vulnerability creeping into him. When the system began recording his brain waves, Paolo felt, for a moment, as if a part of him—an invisible, private part—was about to be exposed. An area of his psyche that usually remained hidden beneath the veil of rationality was now becoming the subject of study.

"Let's see what you can do," he murmured to himself, as if he were facing a calculating machine rather than a device that would challenge his understanding of the intangible. The screen displayed a moving graph, a series of lines reflecting his mind's activity. The initial data showed his neutral state, but the app suggested an exercise: close his eyes and imagine a moment of calm. Paolo chose to imagine the sea—a vague memory of a childhood vacation, a remote corner of his mind that he recalled as a place of serenity and escape from the frenetic pace of daily life.

The waves on the graph began to shift, and Paolo watched intently as the app showed an increase in alpha waves, an indicator of relaxation. He felt surprised and fascinated. His mind, which usually moved through patterns and calculations, was now producing something that could be measured and visualized. *"Is this really working?"* he wondered, his tone almost incredulous. He couldn't believe that his mind, which he had always treated as a set of logical responses and solutions, could be influenced by images evoked in his imagination and monitored by a simple sensor.

He spent the entire afternoon experimenting. He tried imagining different moments: the satisfaction of solving a problem at work, the joy of a past achievement, the warmth of companionship among friends. Each time, the graph responded, changing its values and signals, and Paolo couldn't help but feel intrigued. It was as if the device was telling a story he himself didn't know—a language made of numbers and waves that represented emotions he had always considered too fleeting to be captured.

When he finally turned everything off and removed the headband, he found himself staring at the table, lost in thought. The silence of the room seemed to amplify his state of mind. *"Maybe there really is more to this than I can measure,"* he murmured, a faint, uncertain smile on his lips. That first contact with the part of his mind he had never been able to define had opened a door that, until now, had remained closed. Maybe there really was an invisible world—a world of emotions and thoughts—that deserved to be explored.

Chapter 4:

Francesca - Inside the Imagination

✦ ✦ ✦

Francesca carefully settled onto the soft rug in her favorite room, her heart beating slightly faster with excitement. She was surrounded by cushions, each in a warm and inviting color, and the dim light of a lamp created an intimate and reassuring atmosphere. The room, with its serene walls and the scent of old books, seemed like the perfect place to dive deep into her mind. She had been eagerly awaiting this moment—the chance to delve even further into her inner world, to explore it without limits, guided by that small device that promised to give visible form to her thoughts, as if she had finally found a bridge between her spirit and technology.

She placed the **eMind EEG** band on her head with care, as if preparing her body and mind for a sacred ritual. The sensation of the sensor resting gently on her forehead made her smile, as though the device were the guardian of a hidden potential waiting to be discovered. A slight vibration on her smartphone signaled that the system was ready. Taking a deep breath and pausing for a moment of silence, Francesca closed her eyes, letting her mind wander freely, without forcing any thoughts—simply observing them as they unfolded in the space of boundless imagination.

The first image that emerged in her mind was of a green meadow, a place she had often imagined as a child. The scene materialized vividly: the deep blue of the sky, the white clouds as soft as cotton candy, as if she could reach out and touch them. She could almost feel the warmth of the sun on her skin and the rustling of grass beneath her bare feet. Every element, every sound and sensation, was so vivid it felt real.

The device began recording, and Francesca opened one eye to observe the graph forming on the tablet screen. The **alpha waves** started to rise, indicating a state of deep relaxation. A smile appeared on her lips. It was as if the device was confirming what she had always known: imagination was her refuge, her safe space, a realm without boundaries, where no limitations of reality applied.

She decided to push further, to explore even more. She imagined herself beneath a large tree, its green leaves filtering the sunlight into golden rays that danced on the ground. It was a place that felt familiar, even though it didn't physically exist. Beneath it, a bench sat empty, inviting her to rest. She sat down, allowing the deep sense of peace she felt to expand throughout her body, like a warm blanket wrapping around her gently. The graph on the screen started showing different peaks, a sign that her emotions were shifting. Every thought seemed to amplify her connection with that imaginary space.

"What would happen if I imagined a hug?" she wondered, her heart beating a little faster, as if the question had stirred something deep within her. It was a scene she often evoked in

moments of solitude, and it never failed to make her feel better. The image in her mind took shape: a figure approaching, its face indistinct yet warm and friendly. When the embrace happened, the graph showed a sudden increase in **gamma waves**.

Francesca's heart filled with joy—an emotion she couldn't fully explain but felt intensely. The device wasn't just recording data; it was confirming that imagination could create a tangible emotional reality—one just as real as the physical world.

When she finally removed the headband, she felt a calm she had rarely experienced before. Her mind was at peace, as if a weight had been lifted. The experience had taken her to a level of connection with herself that she had never thought possible. She sat for a moment, letting the inner peace settle completely.

"There's so much to explore," she whispered, holding the device in her hands as if it were a precious object, a key unlocking a new dimension of her being.

That evening, she wrote in her journal:

"Imagination is not just a refuge. It's an entire world, a place where we can be anything we want."

The words lingered on the page, carrying a meaning she had never felt before, as though she had just discovered a new horizon within herself—one she would continue to explore.

Chapter 5:

Paolo - The First Discovery

✦ ✦ ✦

Paolo sat in his study, his gaze fixed on the device resting on the table, as if it were a mysterious object—a key to a world he did not yet fully understand. It had been a few days since he had tried it for the first time, and he couldn't stop thinking about what he had seen. The graphs, the brain waves shifting with his thoughts—it was as if a window had opened onto a part of himself he had always ignored. A dimension he could neither measure nor analyze like a work problem, yet one that carried just as much weight.

He decided to push further. This time, he wanted to conduct a more targeted experiment, something that would test not only the device but also his own mind. It was a step outside his comfort zone, but Paolo felt that the moment had come to explore this new frontier—to challenge the rationality that had always governed his life.

Seated at his desk, he put on the headband and launched the app, preparing to face an experience he had not yet fully accepted.

"Imagine a scene of personal success," the app suggested.

Paolo closed his eyes, letting the silence of the room envelop him, like a gentle touch inviting him to surrender. He searched for a memory that had a strong impact on him, a moment engraved in his mind like a milestone. His mind returned to a corporate award ceremony from years ago, when his team had won an important recognition. He could still hear the applause, feel the weight of the trophy in his hands, see the satisfied smile of his boss. Every detail, every emotion from that moment, resurfaced with clarity.

The brain wave graph displayed an increase in **beta waves**, indicating intense cognitive activity and concentration. Paolo opened his eyes briefly, observing the data with interest. He was truly reliving that emotion in his mind, as if the device were tracing an invisible but tangible path. The graph didn't lie—his body and mind were reacting.

But the real discovery came when he tried something different. He wondered what would happen if he imagined something that had never actually happened.

An idea struck him like a spark illuminating a dark corner of his mind: imagining a moment of emotional connection, something beyond work or tangible achievements, something that spoke directly to his heart.

He closed his eyes again and forced himself to visualize a scene he had never experienced but somehow felt close to. He imagined himself by a lake at sunset, with someone beside him. He couldn't define the features of this person, but the sensation was clear—a feeling of peace, of belonging, as if all his worries had disappeared. It was a scene he couldn't fully explain, yet it made him feel in tune with the universe.

The graph showed a spike in **gamma waves**, a sign of heightened emotional activity. Paolo remained still, letting the image envelop him completely. His heart beat faster, and a wave of warmth washed over him.

When he opened his eyes, he realized that something inside him had changed—something he couldn't quite put into words.

"It's just imagination," he told himself, trying to rationalize it. But deep down, he knew that this scene, though not real, had touched something profound.

The mind was more powerful than he had ever believed.

That first encounter with imagination had shown him a side of his psyche he had never acknowledged before, and his curiosity about what he might discover in the future grew exponentially.

Chapter 6:

Francesca - The Room of Memories

✦ ✦ ✦

Francesca loved how imagination could take her back to moments in her life she thought she had forgotten. It was as if every memory, every emotion, was stored in a hidden corner of her mind, waiting to be rediscovered when needed. That evening, she felt the urge to go deeper—to step into her memories with newfound awareness.

She decided to dedicate herself to a special exercise: exploring her memories through **eMind EEG**, as if embarking on an inward journey to uncover the secrets hidden in her heart.

Seated on the couch, a cup of warm tea in her hands, her body relaxed, but her mind was ready for what was to come. She put on the headband and focused. The app suggested choosing a happy memory—one that had made her feel loved and alive.

Her mind took her back to a summer afternoon, when she was a child playing in her grandmother's garden. She could still smell the roses, feel the sun's warmth on her shoulders, hear her grandmother's crystalline laughter as she told stories—stories that felt timeless.

The graph displayed a gradual increase in **alpha waves**, indicating relaxation and tranquility. Francesca let herself be fully immersed in that moment, losing herself in the details of the memory as if it were a tangible reality. Her heart beat slower, and her body relaxed even further. She felt safe, as though the past had come back to life before her eyes.

But there was something she wanted to understand better.

She decided to evoke a different memory—one that was painful. Her mind, accustomed to navigating serene memories, was not afraid to face the difficult ones as well. She wanted to see how her brain would react, how the device would capture her deepest emotions.

The scene that formed in her mind was one of farewell—a moment she had always avoided thinking about. She saw herself as a young woman, standing in the rain, watching a train leave, carrying away a part of her that would never return. The emotions were so intense that they felt almost real, as if the past had resurfaced with all its strength.

The app registered a spike in **gamma waves**, and Francesca noticed how her body reacted—her heartbeat quickened, and a tear rolled down her cheek. She couldn't stop it, as if it were her body's natural response to a sorrow that had always been there, buried deep inside.

When she reopened her eyes, she took a deep breath.

The device had captured something she had always known but had never fully explored: memories, even painful ones, hold immense power.

But at that moment, she realized that through imagination, she could transform those memories—give them new meaning, as if her heart had found a way to heal, to rewrite her story.

Francesca turned off the device and wrote in her journal:

"The mind does not forget, but it can heal. Through imagination, we can rewrite our story."

The words lingered on the page, but Francesca didn't see them as mere words.

They were a promise to herself—a new vision of the world and of who she was.

Chapter 7:

Paolo - The Mind That Surprises

✦ ✦ ✦

Paolo found himself, for the third consecutive evening, sitting at his desk with the **eMind EEG** already positioned on his head. Every time he prepared to put on the device, he felt a strange sense of anticipation, as if he were about to embark on a journey into unknown territory.

He had begun to see this moment as an appointment with his own mind—an opportunity to discover what lay behind his usual rational approach. The device, which had initially seemed like just a technological gimmick, was becoming his companion on a journey of self-discovery. A door was opening into a more complex dimension of his psyche—a dimension he could not control with the same precision he used in his work and daily decisions.

That evening, he decided to raise the stakes.

After reading some guides and suggestions on how to use the device, he had come across an advanced exercise: imagining something entirely new—something beyond the boundaries of his experiences. A challenge to his rational mind, so accustomed to control.

With a slight smirk of challenge, Paolo closed his eyes.

He let his mind wander for a few moments, trying to clear his thoughts, then began constructing an image.

He was on a snow-covered mountain, its peak shrouded in clouds. The air was crisp and biting, and he could feel the cold on his face. He wore a heavy jacket, and the sound of his footsteps crunching into the fresh snow was muffled. In his hands, he held a compass, as if searching for direction—as if the mountain were a metaphor for the inner journey he was undertaking.

Every detail was vivid: the icy wind tousling his hair, the sound of his breath growing deeper with each step.

There was an adrenaline rush in the air—the thrill of being there, just steps from the summit, feeling as though he were challenging both himself and nature at the same time.

The graph on the screen reacted, and the **gamma waves** showed intense activity.

Paolo couldn't resist opening one eye to observe the data moving in real-time. The response of his brain fascinated him. He had never lived this scene, yet his mind was treating it as if it were real.

"And if I add someone else?" he wondered.

The solitude of the mountain now felt more intense, as if the device had sparked a deeper reflection on human connection.

In his imagination, another figure appeared beside him.

It was an unfamiliar face, but there was something reassuring in their presence.

They walked together toward the summit, their words indistinct yet comforting, as if they were sharing a profound connection without the need for speech.

When he finally removed the headband, Paolo stared at the screen for a few moments.

That imaginative journey had left him with more questions than answers.

He realized that the mind was capable of creating experiences so vivid that they could challenge reality itself.

The line between what was real and what was imagined was becoming increasingly thin.

He wrote a note on a piece of paper next to the device:

"It's not just about imagining. It's about living what hasn't happened yet."

Chapter 8:

Francesca - Beyond the Boundary

✦ ✦ ✦

Francesca lit a scented candle, its sweet and calming aroma filling the air, and sat cross-legged on the rug.

That evening, she felt the need to push further—to explore something she had never dared to imagine before, to step beyond her comfort zone, to discover new corners of her mind.

The usual tranquility of her home, the serenity of her room, was no longer enough.

She felt an urge to face something deeper—something greater.

She placed the **eMind EEG** band on her head and launched the app, but this time, she decided to ignore the guided exercises.

She didn't want to follow a preset script; she didn't want to be confined by predefined patterns.

She wanted to let her mind flow freely, letting her thoughts run wild without restrictions.

That evening, she wanted to be completely authentic with herself—without external influence.

Closing her eyes, she found herself in an unfamiliar place: a deserted beach at sunset.

The sky was painted in shades of orange and pink, while the calm sea stretched to the horizon, a vast mirror reflecting the colors above.

Francesca walked along the shore, letting the waves wash over her feet.

Every step invited her to sink deeper into that feeling of peace and freedom that only the ocean could give her.

The scent of salt filled the air, and the rhythmic sound of the waves was the only melody accompanying her.

Then, a noise caught her attention.

She turned and saw a small sailboat anchored near the shore.

Without hesitation, she imagined herself stepping onto it, setting sail.

The feeling of the wind in her hair, the movement of the water beneath her—it was so real that it almost took her breath away.

She felt weightless, as if she could fly, as if she were letting go of every thought, every worry, simply drifting on the water.

The graph on the screen showed a surge in **alpha and gamma waves**.

Francesca felt completely immersed in the experience.

There were no boundaries between her mind and the reality she was creating.

There was no longer a difference between what was imagined and what was lived.

But she didn't stop there.

She closed her eyes even tighter and imagined she wasn't alone.

A figure appeared beside her, sitting at the other end of the boat.

She couldn't clearly see the face, but she felt a deep connection, as if that person were a part of her.

Words weren't needed—the silence they shared was full of meaning.

When she reopened her eyes, Francesca felt her heart beating fast.

She removed the device and sat in silence for a few moments.

That journey into imagination had taken her beyond the boundaries of what she knew.

It revealed to her that imagination was not just a refuge but a bridge to something greater.

She had found a new way to connect with herself and with others—one that went beyond words and physical actions.

She wrote in her journal:

"Within us lies an entire universe, and imagination is the key to exploring it."

Every word she wrote felt like a truth she had just uncovered—a newfound awareness of herself.

Chapter 9:

Paolo - The Challenge of Control

✦ ✦ ✦

Paolo had always had a particular relationship with the concept of control.

His life was governed by schedules, rules, and logic.

Everything had to be planned down to the smallest detail.

His approach to life was like a well-oiled machine, where every part had its place and where the unexpected had no room.

But with **eMind EEG**, he had discovered a realm where control seemed to slip away: **his mind**.

That part of himself that eluded his grasp.

That he couldn't measure or premeditate.

Seated at his desk, the device ready, the app already running, Paolo had set himself a clear goal for that evening:

To test if he could control his emotions through imagination.

The challenge intrigued him, and in some ways, unsettled him.

He was about to confront a part of himself he had only observed from a distance.

The control he had always relied on was about to be tested by something he couldn't fully rationalize.

He put on the **eMind EEG** and selected an exercise that required him to **evoke specific emotions**.

The first step was simple: **relaxation**.

Paolo closed his eyes and imagined himself surrounded by the stillness of a forest.

The sound of the wind through the trees and the chirping of birds filled the air, while a gentle sense of calm settled over him.

The graph showed an increase in **alpha waves**, indicating that he was successfully achieving a state of relaxation.

"So far, so good," he thought, almost amused by the simplicity of it.

But the next step was more complex.

He had to imagine **a situation of controlled tension**—something that stimulated his concentration **without triggering anxiety**.

Paolo chose a familiar image: the conference room at his company, the board of directors waiting for his presentation.

The tension was **palpable**, but it didn't overwhelm him.

The graph displayed an increase in **beta waves**, indicating intense cognitive activity.

Paolo felt exactly as he did during an actual meeting: heart slightly accelerated; mind fully focused.

But this time, he knew he was in control.

A single change in perspective could break the tension.

"Now let's try something harder," he murmured, as if daring himself.

The app suggested evoking a memory that triggered **happiness**.

Paolo concentrated, searching for something significant.

His mind took him back to a childhood afternoon, playing soccer with his friends in the backyard.

The sun shone brightly, laughter filled the air, and every detail felt vivid—almost as if it had happened just yesterday.

The **gamma waves** spiked, showing that his mind was entering an intense state of emotional activity.

Paolo realized he was smiling—completely unconsciously.

His mind had responded instinctively to the joyful memory.

But the moment he became aware of that happiness, the scene started to fade.

It was as if his **rational mind** had interrupted the natural flow of imagination.

"It's not as easy as it seems," he admitted, removing the device and setting it on the table.

The control he had always taken for granted was elusive.

His mind wasn't as easily manipulated as numbers and graphs.

Yet, the challenge fascinated him.

That evening, he noted in his journal:

"The mind is like a river. We can redirect its course, but we can never stop its flow entirely."

Chapter 10:

Francesca - The Dance of Emotions

✦ ✦ ✦

Francesca loved to dance.

It wasn't something she did often, but every time she did, she felt her **body and mind merge into a state of perfect harmony**.

It was as if movement itself unlocked something deep inside her, as if dancing allowed her soul to speak.

That evening, she decided to incorporate **eMind EEG** into her experience, curious to see how her brain responded to an activity that engaged both **the physical and the emotional**.

It felt like the right step—blending her love for dance with her desire to explore the potential of her mind.

She pushed the coffee table aside, creating space in the middle of the room.

The polished floor reflected the glow of the lamp, casting a warm, inviting atmosphere.

She chose a playlist of contemporary classical music—melodies capable of transporting her to another dimension.

She placed the device on her head and closed her eyes for a few moments, breathing deeply before she began.

As soon as the music filled the air, she felt **weightless, timeless**.

Every movement was fluid, spontaneous, as if the melody were a language her body had always understood.

She didn't have to think—every step felt natural, inevitable.

Her body **moved instinctively with the rhythm**.

On the screen, her **brain waves reacted in a unique way**:

Alpha waves increased, indicating deep relaxation.

Gamma waves surged, showing intense emotional activity.

Francesca smiled as she watched the graph move—but it wasn't just the science that fascinated her.

It was the experience itself.

The **body in motion**, the **rise and fall of emotions**, mirroring the music's own flow.

At one point, Francesca imagined **she wasn't alone**.

She visualized a **grand ballroom**, filled with other dancers, each lost in their own world, yet moving together in a shared rhythm.

The scene became increasingly **vivid**—

The sound of shoes gliding over the polished floor.

The soft murmur of distant conversations.

The golden glow of warm lights illuminating the space.

The **graph changed again**, showing a perfect synchronization between **emotion and imagination**.

Francesca felt **light, free**—as if she were floating between reality and fantasy.

As if there were no longer a separation between the **body that danced** and the **mind that created**.

It wasn't just about movement.

It was about **the power of creating a reality within the mind and living it through the body**.

When the music ended, Francesca collapsed onto the couch, her breath steady, a serene smile still on her lips.

"It's as if I danced inside and outside of myself at the same time," she thought.

She felt **renewed**—as if she had experienced something profound, a journey she could never take with just her body.

She removed the headband and looked at the recorded data.

The **gamma waves were at the highest levels she had ever seen**.

She opened her journal and wrote:

"Imagination is my dance partner. It takes me where my body alone could never go."

Every word felt like a truth she had just discovered—a **new awareness** of herself.

Chapter 11:

Paolo - The Paradox of the Mind

✦ ✦ ✦

Paolo stared at the graph flashing on the screen, his eyes following every movement of the lines as if he were trying to uncover a hidden meaning in that sea of data. It had almost become a nightly ritual: putting on the device, following the app's instructions, and discovering something new about himself—like an explorer venturing into unknown territory.

But that night, a thought kept lingering in his mind.

"Is it possible that the mind can deceive itself?" he wondered, absentmindedly playing with the pen in his hand while the device rested gently on his head. It was a thought that intrigued him, yet unsettled him at the same time.

He had read an article about the power of suggestion and how imagination could manipulate perceptions, creating realities with no physical basis. He decided to put that theory to the test. The control he had always exercised over his emotions and thoughts was about to be challenged in a way he had never imagined.

Paolo knew that to better understand his mind, he had to uncover its deepest truths.

He put on the headband, set up the device, and prepared to explore. Closing his eyes, he started with a simple, familiar image: a cup of hot coffee in his hands. The scene was clear in his mind—he could almost feel the warmth radiating from the cup and the rich aroma filling the air. Every detail was vivid, as if imagination were a field Paolo knew well.

The graph showed a steady response, with **alpha and beta waves** reflecting relaxation and focus.

"So far, so easy," he thought. It was a simple image, but enough to test the device.

But Paolo wasn't satisfied. The real challenge was yet to come.

He decided to push further. He tried to imagine something he had never seen before, something he couldn't verify—a **smell he had never experienced**.

His mind, usually precise and analytical, began to construct an **imaginary flower** with emerald-green petals and a unique, indescribable scent. Almost effortlessly, his mind created an olfactory sensation so vivid that Paolo was startled. He hadn't just imagined a visual image—he had crafted an entire **sensory experience** as if the flower were truly there with him.

The graph suddenly shifted. **Gamma waves** began to oscillate unpredictably, signaling intense emotional and cognitive activity.

Paolo opened his eyes, staring at the data on the screen, amazed at what he had just experienced.

Had he really managed to create something so powerful in his mind that it felt real?

His mind was capable of simulating experiences with such intensity that they seemed tangible.

"But how much of this is imagination, and how much is real?" he asked himself, removing the device and placing it on the table.

The paradox of the mind fascinated him: it could simulate experiences, yet those simulations could evoke **authentic emotions**, blurring the lines between imagined and lived reality.

That night, he wrote a note in his journal:

"Imagination is not an illusion. It is a reality that lives within us, governed by rules we do not yet fully understand."

Chapter 12:

Francesca - The Voice of the Mind

✦ ✦ ✦

Francesca lay on her bed, surrounded by the silence of her room. It had been an intense day, yet she felt something calling her.

Imagination was not just a refuge for her—it was a doorway to something deeper, a language her mind spoke when she wasn't trying to control it. She sensed that through **eMind EEG**, she could venture even further into that invisible and inexplicable world that was her inner self.

She placed the headband on her forehead and started the app, but this time with a precise goal: she wanted to **listen to her mind**.

She didn't want to create images or complex scenarios; she simply wanted to let her thoughts emerge spontaneously, like a free-flowing current, unforced and unrestricted.

Closing her eyes, she focused on her breath, trying to clear her mind of surface-level thoughts, allowing something deeper to rise.

The first image that surfaced was a **country house**, the place where she had spent summers as a child.

It wasn't a vivid memory, but a sensation—the sound of the wind through the trees, the creak of the rusty gate, the warmth of the sun on her skin.

Francesca felt as though she were walking through her grandmother's garden again, breathing in the fresh air. But her mind didn't keep her there.

It was just the **beginning of a journey** that would take her far beyond.

Then, something changed.

It was no longer an image but a **voice**—not a real voice, but the voice of her subconscious, speaking to her through **emotions and sensations**.

Francesca felt light, as if she were floating between memories and thoughts that had never fully surfaced. Her mind seemed to speak a language she didn't consciously know, yet deeply understood.

Every thought that emerged carried meaning, like reading the pages of a book she had written without realizing it.

The **gamma waves** on the graph spiked, indicating that her brain was processing something profound.

Francesca remained in that state for several minutes, allowing the images and sensations to pass through her without judgment, like a river gently carrying her along.

It was as if she were learning a **new language**—a language that spoke directly to her soul.

When she opened her eyes, she felt as though she had **conversed with a part of herself she had always ignored**.

It wasn't an encounter with something external but with a hidden dimension of herself that was now coming to light.

She removed the headband and lay still, staring at the ceiling with a serene smile.

"The mind doesn't speak in words," she thought, *"it speaks in emotions."*

It was a language she had never truly understood before, but now it felt clearer.

She wrote in her journal:

"I don't need to seek answers outside myself. My mind already knows the questions and the answers—if only I am willing to listen."

Those words carried a **special weight**.

They were no longer just reflections, but a **deep realization growing within her**.

Chapter 13:

PaoloA Journey into the Unknown

✦ ✦ ✦

Paolo sat at his desk, the device ready, like an explorer preparing for a new adventure.

Every session with **eMind EEG** left him with more **questions than answers**, but that night, he felt the need to go further.

It was as if the **boundary between what he knew and what he had never dared to explore** had become thinner.

He wanted to see **how far his mind could go**, to explore territories he had never considered, to **discover dimensions of his psyche that even he didn't know existed**.

He started the app and closed his eyes.

This time, he would follow no guides or instructions.

He decided to **let his imagination take him wherever it wanted—no plans, no expectations**.

He felt as if he were about to **cross a threshold** without knowing where it would lead.

At first, nothing happened.

The darkness behind his closed eyelids was still, almost monotonous.

For a moment, Paolo wondered if he was truly succeeding in freeing his mind, or if he was just trying to **force something that wasn't meant to surface**.

Then, **slowly, a scene began to emerge**.

He found himself in a **mysterious city**, with **narrow cobblestone streets**, lit by streetlamps casting a warm glow.

The houses were old, with **balconies overflowing with flowers** and **windows open**, letting fragments of music and laughter spill into the air.

The scene was **so vivid** that Paolo felt strangely **at home**, as if he had been there **in another life**, as if the city had **a deep connection to his soul**.

Every step he took down those streets seemed to confirm that feeling of familiarity—as if **time held no meaning in that place**.

The **gamma waves** on the screen spiked, signaling that his brain was actively constructing the place **in real-time**, adding **details with every step**, making them **increasingly real**.

Paolo stopped in front of a **small square**, at the center of which stood a **fountain**, its water flowing gently, creating a **delicate melody**.

He stepped closer and **knelt to touch the water**.

Even though he **knew** it was just **imagination**, the **sense of reality was so strong that he shivered**.

The coolness, the shimmering water under the streetlamp—it all **felt tangible**.

Then, something **unexpected** happened.

A **figure appeared** on the other side of the square.

It was **indistinct**, wrapped in a **soft glow**, but Paolo felt a **strange connection** to its presence.

His **mind, which had so accurately crafted the city,** now seemed intent on **completing the picture with this mysterious figure**.

He tried to **focus**, to **see more clearly**, but **the figure vanished into nothingness**, leaving him with a feeling of **inexplicable loss**—as if he had caught a glimpse of something **fundamental** but had failed to reach it.

His eyes snapped open, his heart pounding, as if that **moment of connection had left an emptiness inside him.**

The graph showed **a sudden peak**, as if his mind had reached **an extraordinarily high level of activity**.

He removed the device and leaned back in his chair, trying to **rationalize the experience**.

But **something inside him resisted logic**.

"It wasn't real," he told himself.

But deep down, he **knew that wasn't enough to explain what he had felt.**

His mind had created **something so powerful** that it seemed to go **beyond imagination itself**.

He wrote in his journal:

"Imagination is not just creation. It is a bridge to something we do not yet understand."

Chapter 14:

Francesca - The Invisible Embrace

✦ ✦ ✦

Francesca sat on the carpet, the soft glow of the lamp casting gentle shadows on the walls—an atmosphere that invited reflection.

That evening, she felt a deep need for **connection**, for exploring an **emotion she had been missing for a long time**: the feeling of an **embrace**, of human warmth that seemed distant from her everyday life.

It wasn't just **a physical sensation** she was searching for, but something **deeper**—an emotional encounter that transcended touch, allowing her to **feel again** that reassuring, warm presence.

She placed the **eMind EEG** device on her head and closed her eyes, allowing her mind to open **like a flower blooming at dawn**.

The app suggested **evoking a happy memory**, but Francesca **wanted more**.

She didn't just want to **remember**—she wanted to **create something new**, something that could **nourish her soul**.

This wasn't just about **reliving the past**—it was about **bringing to life an experience that was unique, personal, and spoke directly to her deepest desires**.

She imagined herself in a **cozy room**, with warm walls and a fireplace casting a soft, golden glow.

She stood in the **center of the room**, wrapped in a **feeling of anticipation**.

It was as if **something was about to happen**, and she simply had to **wait**.

Then, slowly, she **imagined a figure approaching**.

She couldn't see its face clearly, but she **felt its presence—reassuring, enveloping**.

The figure seemed to **make its way into her mind**, as if it had **always been a part of her**, but never **so tangible** before.

When she imagined the **embrace**, the **gamma waves** on the screen spiked—indicating intense **emotional activity**.

Francesca **felt it happening**—the **warmth of arms around her**, the steady rhythm of a **heartbeat against her own**.

There were **no words**, only a **deep peace** that **filled every empty space within her**.

It was as if **her imagination had forged a connection**—not just with someone else, but with **herself**.

A bond that, though **without physical form**, felt **more powerful than any tangible reality**.

When she opened her eyes, a **tear slipped down her cheek**—but it wasn't **sadness**.

It was **gratitude**.

She was no longer **alone**—there was only **a sense of wholeness** that her imagination had offered her.

She had **found what she was looking for**, and she had found it **within herself**.

Imagination had given voice to her **need for connection**—an **invisible embrace** that had never needed **to be physically expressed**.

She removed the device and wrote in her journal:

"Sometimes, what we need is not outside of us but within. Imagination can give us what the world denies us."

The words she wrote felt **heavy with truth**, as if she had **just uncovered a new part of herself**, a part she was **finally ready to explore**.

Chapter 15:

Paolo - The Threshold of Reality

✦✦✦

Paolo stared at the device on the table **as if it were a puzzle waiting to be solved.**

Every session with **eMind EEG** left him with **more questions than answers**, but tonight, he felt the **urge to push further**.

The journey he had begun **into his own mind** was becoming **increasingly complex and fascinating**, and his curiosity **compelled him** to test the **boundaries between imagination and reality.**

He no longer wanted to simply **explore**—he wanted to **understand**.

He wanted to see if **the mind was truly capable of transcending the barriers of perception and physical reality**.

He placed the headband on his forehead, launched the app, and closed his eyes.

This time, he wouldn't follow **any pre-set exercises**.

He didn't want to be **tied down to structured instructions**.

He just wanted to **let his imagination take him wherever it wanted—without restrictions**.

The mind had become, for him, a kind of **unmapped land**—where **every step was a discovery**, **every thought a new possibility**.

At first, **nothing happened**.

The **darkness behind his closed eyes was motionless, almost empty**.

Paolo wondered if it was **possible to truly empty the mind**, or if he was simply **forcing something that wasn't meant to happen**.

Then, **slowly, a scene began to take shape**.

He found himself in a **mysterious city**, its **narrow streets paved with old stones, bathed in warm lamplight**.

The buildings were **ancient**, adorned with **balconies full of flowers**, and **music and laughter drifted from open windows**.

The scene was **so vividly real** that Paolo felt **a strange sense of belonging**, as if he had **been there before**.

The **gamma waves** on the screen surged, indicating that his brain was actively **constructing the environment in real time**, adding details with **each step**.

He walked down the streets, feeling **both a deep connection and a sense of mystery**.

Then, he arrived at a **small square** with a **fountain at its center**, water trickling softly, creating a **delicate melody**.

He knelt beside the fountain and **reached out to touch the water**.

Even though he **knew** it was just **imagination**, the **sensation was so real that he shivered.**

The **coolness of the water**, the **reflection of the lamplight—it all felt tangible.**

Then, **something unexpected happened.**

A **figure appeared** on the **other side of the square.**

It was **shrouded in soft light**, **indistinct**, but Paolo felt **an inexplicable connection.**

His **mind, which had created the city with such precision,** now seemed to be **weaving this figure into the experience as well.**

He **tried to focus, tried to see it more clearly**—but before he could, **the figure vanished into nothingness.**

He was left with a feeling of **unexplainable loss**, as if he had **almost reached something important but couldn't grasp it.**

His eyes flew open, his **heartbeat racing**, as if that **brief moment of connection had left a void inside him.**

The graph displayed a **sudden peak**, indicating an **extraordinarily high level of mental activity.**

Paolo removed the device and leaned back, trying to **rationalize what had just happened.**

But **something inside him resisted logic.**

"It wasn't real," he told himself.

But deep down, he **knew that wasn't enough to explain what he had felt.**

His mind had **created something so powerful** that it seemed to **transcend the boundaries of imagination.**

He wrote in his journal:

"There is a threshold between imagination and reality. Perhaps they are not as far apart as I once thought."

Chapter 16:

Francesca - The Symphony of the Mind

✦ ✦ ✦

Francesca turned on her **favorite playlist** and sat cross-legged on the carpet, as if preparing for a journey—**a journey far from reality**.

That evening, she felt that **music would be the key** to unlocking **new doors in her imagination**.

Music had always held a **special power** over her, but this time, she wanted to **see how powerful it could become when combined with eMind EEG**.

She wanted to **explore how much her mind could create**—how many **complex worlds** she could build using **nothing but sound and imagination**.

She placed the headband on her forehead and let the **first track** fill the room.

It was **a soft, melancholic melody**, the kind of music that **gets under your skin** and makes you **travel far away**.

Francesca closed her eyes, **feeling every note like a gentle touch**.

There, in the **silence of her room**, the **music became a language**—one that only **her mind could understand**.

She imagined herself standing in **an empty theater**, at the **center of the stage**, with **dim lights illuminating only her**.

The **solitude of the place**, combined with the **power of the music**, made her feel more **alive than ever before**.

The music **changed**, becoming **livelier**, and in her mind, **the theater came to life**.

An **invisible orchestra** played a **symphony that seemed to emerge directly from her emotions**.

Every instrument represented a **different feeling**:

The **violin** for **nostalgia**

The **flute** for **joy**

The **piano** for **hope**

The melody grew more **complex**, as if her mind were **writing a story as she listened**.

Francesca realized that her mind was not just **listening** to the music—it was also **creating it**.

She was the **conductor of this invisible orchestra**, and with **every movement of her hands**, the melody **shifted**—becoming **more powerful, more alive**.

She wasn't just **imagining**.

She was **composing**, giving life to a **symphony of the mind**—a symphony that **expressed everything she carried inside her**.

The **brainwave graph** recorded intense activity.

The **gamma waves** peaked, showing that Francesca was experiencing a **deep emotional state**.

But the **true magic happened** when she imagined **herself conducting the orchestra**.

She raised her hands, and with every **gesture**, the music **transformed**.

She wasn't just **following the melody**—she was **creating it**.

When the music **finally stopped**, Francesca opened her eyes, still **immersed in a state of awe**.

The device had recorded every moment, but **what truly mattered** was **not the data**.

It was **the journey she had taken**, the realization that **her mind could create entire worlds—worlds that didn't need an audience to exist**.

This was **her own world**, made of **emotions and sounds**, and for the **first time**, she felt like she had **complete control over it**.

She wrote in her journal:

"Imagination is my orchestra, and every emotion is a note. We can compose symphonies that resonate within us—if only we have the courage to listen."

Chapter 17:

Paolo - The Power of Memory

✦ ✦ ✦

Paolo stared at the **device**, watching as the **blue light pulsed**, signaling that it was ready to be used.

That night, he had **no specific plan**, but something was **bothering him.**

For days, a **memory had been haunting him**—a memory that seemed to **surface from the depths of his mind**, like **a wound that had never fully healed.**

He couldn't shake it off, and he knew that **to move forward, he had to confront it.**

He placed the headband on his forehead, launched the app, and closed his eyes.

This time, he didn't want to **imagine something new.**

He didn't want to **experiment with creativity.**

He simply wanted to **face the memory that made him uncomfortable**—something that had **always escaped him** but that he knew was there, waiting to be explored.

Taking a deep breath, he let his mind **carry him back in time**.

It was an **autumn evening, many years ago**.

Paolo saw himself sitting on a **bench in an empty park**, with leaves **falling lightly around him**.

He had just ended an **important relationship**, and the **weight of that decision pressed against his chest**.

He remembered the **cold air** seeping through his coat and the **silence** that seemed to **scream his loneliness**.

The world around him felt **gray and lifeless**, and **every breath felt heavy**—as if he had lost **something he could never get back**.

As he **relived that moment**, the **gamma waves** on the screen spiked, signaling that **his emotions had reached an intense level**.

Paolo felt **vulnerable**, but he didn't **run away**.

He decided to **stay with that memory**, to **stop avoiding it** as he had done so many times before.

This time, he wanted to **face it, analyze it**, and perhaps—**rewrite it**.

With a conscious effort, he **imagined himself sitting on that same bench again**.

But this time, **he was not alone**.

Next to him, there was a **presence**—not a **defined face**, but a **figure that radiated warmth and understanding**.

Slowly, the **scene began to change**.

It was no longer a **moment of pain and isolation**, but an **encounter of comfort**.

Paolo felt as if he were **healing a part of himself**—a part that he had **always avoided facing**.

The **presence beside him** helped him **see the moment from a new perspective**, making the burden feel **lighter**, as if **the pain had been shared** and could finally be **let go**.

When he opened his eyes, his **breathing was slow and steady**, and he felt **a deep sense of peace**.

He looked at the **data on the screen**—the **graph showed a balance** between **gamma and alpha waves**, indicating that his brain had **processed the emotion in a healthier way**.

Paolo felt **lighter**, as if he had **finally given the right space to a painful memory—without letting it consume him**.

He wrote in his journal:

"We cannot change the past, but we can change the way we carry it with us. Imagination is a tool to rewrite what hurts us."

Chapter 18:

Francesca - A New Beginning

✦ ✦ ✦

Francesca sat on her **balcony**, wrapped in a **blanket**, as the **cool evening breeze brushed against her face.**

The air was **crisp and fresh**, carrying with it a sense of **change**, as if the **world around her was shifting.**

That night, she felt an **energy different from usual**—a **desire to let herself be inspired by the unknown.**

She wanted to **imagine a future**—something that **didn't exist yet but could become real.**

Like a **blank canvas, waiting to be painted.**

She placed the **eMind EEG** device on her head and closed her eyes.

Her mind started to build an image:

A **bright room**, with **large windows that allowed the sunlight to flood in**, casting a golden glow on everything inside.

She was **sitting on a comfortable couch**, a **notebook open on her lap**, ready to **write something new**—something she **couldn't yet define**.

Soft **music played in the background**, adding a gentle rhythm to her thoughts.

The **air smelled of coffee**, bringing a **sense of warmth and familiarity**.

Then, in her **imagination**, she saw **a door slowly opening**.

A **figure entered the room**.

It was **someone new**, a **face she didn't recognize**, yet **felt strangely familiar**—as if she had **already met them in a dream**.

Francesca didn't know **who this person was**, but she felt **safe, at peace**.

There was no **rush, no urgency**, just a **flow of serenity filling the space**—as if the **future were already there, waiting to be discovered**.

And as she let her mind **wander deeper**, she realized something **powerful**:

She wasn't just **imagining the future**.

She was **creating it**.

Every **small detail**—the **light**, the **music**, the **warm presence in the room**—was a **projection of her own desires**.

A **glimpse into a life she wanted to live**.

The **brainwave graph** showed **a perfect balance between alpha and gamma waves**, meaning her **mind was harmonizing relaxation with deep emotional exploration**.

When Francesca **opened her eyes**, she felt **a quiet confidence within her**.

That **vision** wasn't real yet, but it had given her **direction, clarity, and a sense of hope.**

For the **first time in a long time**, she felt **ready to embrace the future with open arms.**

She wrote in her journal:

"Imagining the future is not escaping the present. It is a way to remind ourselves that we always have the power to choose who we want to become."

The words felt like a **promise to herself**—a **commitment not to let circumstances define her**, but to take control of her own **destiny.**

Chapter 19:

Paolo - Beyond Logic

✦ ✦ ✦

Paolo had always approached life with logic and precision.

For him, every problem had a solution, and every cause had an effect.

Rationality had always been his guiding principle, the compass that directed his every decision.

But ever since he started using **eMind EEG**, he realized that not everything could be explained through reason alone.

That night, he decided to do something different:

To set aside logic and allow the unpredictable to take over.

No filters.

No plans.

Just pure, unrestricted thought.

He placed the device on his forehead, started the app, and closed his eyes.

This time, he wouldn't guide his thoughts.

He would let his mind wander freely—like a river flowing toward the unknown.

At first, there was only darkness.

A vast, empty void.

Paolo waited.

Then, like a flash of lightning across a clear night sky, fragments of images began to appear:

A stone staircase, leading to nowhere

A sky full of pulsating stars, like distant beating hearts

A whispering wind, speaking in a language he couldn't understand

Each fragment seemed disconnected, yet at the same time, part of a greater mystery.

The gamma waves on the screen surged, indicating that his mind was in full creative mode.

Paolo found himself climbing the stone staircase, step by step, feeling the rough texture beneath his fingers.

Each step brought him closer to a light at the top, a light that seemed to call him, yet refused to reveal its nature.

He didn't know what awaited him, yet he kept climbing, driven by curiosity stronger than fear.

At the summit, he encountered an ancient wooden door, covered in intricate carvings that seemed to tell forgotten stories.

Each symbol seemed alive, vibrating with meaning, though he couldn't decipher them yet.

Hesitantly, he reached out and pushed the door open.

Beyond it, there was no physical space—only a feeling:

A sense of absolute freedom, unlike anything he had ever felt before.

It was as if he had crossed a threshold—not in the world, but within himself.

He felt as though he had let go of something old and made room for something new—something unknown but incredibly powerful.

When he opened his eyes, he felt different.

The brainwave graph showed a perfect mix of gamma and alpha waves—an unusual combination that defied all logic.

A balance between control and surrender.

For the first time, Paolo felt truly alive—as if he had touched a part of himself he had never known existed.

He wrote in his journal:

"Logic takes us far, but it is imagination that takes us beyond."

Chapter 20:

Francesca - The Strength of Desire

✦ ✦ ✦

Francesca woke up with a **persistent thought** in her mind.

There was a **desire** she had always kept **hidden**, something she had **never fully explored**.

A desire that spoke of **possibilities**, of **dreams yet to be realized**.

That evening, she decided to **face it head-on**—to use her **imagination to shape what she felt inside**, to **uncover what was hidden beneath this deep longing**.

She curled up on the **sofa**, wrapped in a **soft blanket**, with the **device ready** and the **app launched**.

Closing her eyes, she let her **mind take her where it wanted**.

The first image that emerged was a **vast garden**, filled with **flowers of every imaginable color**.

The **petals seemed to dance** in rhythm with the gentle **breeze**, and with every breath, Francesca inhaled the **intoxicating fragrance** of the blossoms.

At the center of the garden stood a **fountain**, its water flowing softly, creating a **soothing melody**.

Beside the fountain, a **pathway disappeared into the trees**.

Francesca followed it, **feeling the soft earth beneath her feet**, listening to the **birds singing overhead**.

Every **step** felt like an **invitation**—to venture **deeper**, to **go beyond** the surface of her desires.

As she walked **further into the forest**, she felt a growing **sense of anticipation**, as if she were on the verge of **something important**.

At the end of the path, she reached a **clearing illuminated by moonlight**.

In the center of the clearing was a **large mirror of water**, perfectly still, reflecting the **night sky above**.

Francesca leaned forward and gazed into the **reflection**.

But what she saw was **not just her face**.

She saw **versions of herself**—

The **child who dreamed without limits**

The **teenager full of hope**

The **woman she was today**, carrying **all her experiences, pains, and victories**

Each **version of herself** smiled back at her, reminding her that **every phase of her life had been essential** in bringing her to **this very moment**.

As she watched, **a wave of emotions washed over her**—

Nostalgia for **what had been**,
Gratitude for **what she had**,
And a **profound desire to reconnect** with **every part of herself**.

It was as if the **mirror was not just reflecting her face**, but **her soul**—with all its **layers, contradictions, and truths**.

When Francesca opened her eyes, **a single tear** slipped down her cheek—not from sadness, but from **a deep sense of completeness**.

The **brainwave graph** showed a **perfect balance** between **alpha and gamma waves**, meaning her mind had reached a **state of emotional harmony**.

She felt **lighter, more connected to herself**, as if she had found a **missing piece of her inner puzzle**.

She wrote in her journal:

"Desire is not just about what we want. It is what reminds us of who we truly are."

Those words felt like a **promise**—a vow to **never ignore her desires again**, but to **listen, nurture, and follow them wherever they led**.

Chapter 21:

Paolo - The Weight of Choices

✦ ✦ ✦

Paolo sat in his study, the device in front of him, the blue light pulsing gently—waiting for him.

For days, a thought had been haunting him:

A decision he had made years ago, one that, despite the passage of time, still weighed on him.

The memory was vague, but the feeling of regret never left him—like a shadow following his every step.

He often wondered whether he had made the right choice, but every time, he found himself at an impasse.

That evening, he decided to use his imagination—not to escape—but to face the past and, perhaps, finally find an answer.

He placed the headband on his forehead, launched the app, and closed his eyes.

The first image that emerged was a fork in the road.

Two paths stretched before him:

One was straight, paved, and lined with streetlights—well-lit and predictable.

The other was a dirt path, winding through the shadows of tall trees—uncertain and mysterious.

Paolo knew that this fork represented a real moment in his life, but he wasn't sure which decision it symbolized.

It was as if his mind was turning his past into a metaphor—one that he needed to decipher.

He decided to take the well-lit path.

Every step was safe, predictable, guided by the steady rhythm of his footsteps on the pavement.

As he walked, scenes from his past unfolded around him—

Himself in the office, deeply focused on his work

Endless meetings, signing contracts, achieving milestones

A life of control, precision, and success

But as he walked further, he felt a strange emptiness creeping in.

The landscape became dull and gray, the sounds muted, as if something was missing.

At the end of the road, he reached a large mirror.

He stopped and gazed at his reflection.

It was flawless, composed—but completely devoid of emotion.

It was like looking at a statue, an image of himself without life.

The sight disturbed him.

This was not what he wanted to see.

He turned back and decided to take the other path—the dirt trail that disappeared into the trees.

As soon as he stepped onto it, he felt the uneven ground beneath his feet, the cool shade of the trees, the golden rays of sunlight filtering through the leaves.

He saw different images along this path—

Himself laughing with friends

Exploring new places

Embracing the unexpected

Each step felt more real, more authentic.

At the end of the path, he found another mirror.

This time, his reflection was different.

His face looked the same, but his eyes were full of life—they shone with the intensity of someone who had truly lived.

It was not a perfect image, but it was real.

Paolo removed the device and sat in silence, his heart pounding with realization.

The brainwave graph showed intense activity—with spikes in gamma and alpha waves, indicating a profound emotional processing.

He wrote in his journal:

"Choices are never right or wrong. They are simply different roads—and each one leads us to discover a different part **of ourselves.**"

Chapter 22:

Francesca - A Promise to Herself

✦ ✦ ✦

Francesca sat by the **window**, her gaze lost in the **starry sky.**

The night was **clear**, and the stars seemed to **pulse**, as if they were **dancing just for her**.

That evening, she felt the need to **make a promise to herself**—a promise that would be **hers alone**, an **unbreakable commitment to her future**.

It was a **moment of intimacy with her own soul**, an opportunity to **listen to her deepest desires**.

She placed the **eMind EEG** device on her head and closed her eyes.

Her mind took her to a **hidden clearing**, surrounded by **tall, ancient trees**.

The air was **fresh**, carrying the **scent of moss and damp earth**.

At the **center of the clearing**, there was a **circle of ancient stones**, smoothed by time, as if they had been **waiting for centuries** for someone to arrive.

Francesca stepped into the **stone circle** and sat down, feeling the **solid rock beneath her**.

She closed her eyes, allowing the **silence to embrace her**.

"What do I want for myself?" she asked inwardly.

She didn't **force an answer**—she simply **waited**, letting the **images come naturally**.

She saw herself **painting** in a **sunlit studio**, with **large blank canvases waiting to be filled with colors**.

She saw herself **traveling**, discovering **new cultures**, experiencing **new emotions**.

She saw herself **laughing with people she loved**, feeling **connected, present, alive**.

Each **image** carried an **emotion**: **joy, serenity, curiosity**.

Each **scene** seemed to **whisper to her**—reminding her that **everything she wanted was still possible**.

Then, she saw something that **deeply moved her**.

She imagined herself **on a quiet beach at sunset**, a **book open on her lap**.

But it wasn't just any book—**it was her journal**, filled with her **thoughts, dreams, and promises**.

And in that moment, she **realized what she truly wanted**:

She didn't just want to **exist**.

She wanted to **live with intention**.

She wanted **each day to be meaningful**, to be a **chapter worth writing in her own story**.

When she **opened her eyes**, the **brainwave graph** showed a **stable and harmonious activity**.

Francesca felt **calm**, but also **determined.**

That **vision** was not just a **dream**—it was a **plan**, a **direction for her future**.

She wrote in her journal:

"I promise myself not to live just to exist, but to live to create, love, and discover."

She **closed the journal** and looked back at the **starry sky**, feeling **lighter, freer**.

She knew that **this promise would be the guiding light** she needed in her darkest nights.

Chapter 23:

Paolo - The Balance of Emotions

✦ ✦ ✦

Paolo woke up earlier than usual.

The morning light filtered through the blinds, casting soft shadows across the room.

It was unusual for him to feel so restless so early, as if a lingering thought had kept him awake all night.

He made himself a cup of coffee, but the bitter taste did nothing to quiet the unease inside him.

He decided not to wait until the evening—as he usually did—to use **eMind EEG**.

He felt that this was the moment to face what had been weighing on him.

Sitting at the kitchen table, the coffee now cold beside him, he placed the device on his forehead and started the app.

This time, he didn't want to revisit the past or think about the future.

He needed to make sense of the present—to sort through the emotions that had become louder since he started using the device.

Closing his eyes, he found himself in a white room, empty and vast—a pure, untouched space.

At the center of the room, orbs of light floated in the air, each one glowing in a different color.

They were bright, pulsing, almost as if they were alive.

Paolo walked forward, sensing that each orb represented an emotion:

Red for anger

Blue for calmness

Yellow for happiness

Gray for sadness

Instinctively, he reached for the red orb.

The moment he touched it, he felt a wave of heat rush through him—like fire burning from the inside.

He saw memories of frustration—arguments at work, misunderstandings, moments when he had let his emotions control him.

For the first time, he didn't resist.

He simply observed, letting the anger flow through him without fighting it.

And as he did, the red orb softened, its intensity fading.

As if recognizing the emotion was enough to release some of its power.

Next, he reached for the blue orb.

The moment he touched it, a calmness spread through his body.

The memories that emerged were different:

Hiking on a quiet mountain path

Reading a book under a tree

The sound of water flowing in a small creek

It was a reminder of the moments of peace that he rarely appreciated but deeply needed.

Lastly, he reached for the gray orb.

A cold sensation ran through his chest.

With it came memories of loss, disappointments, and moments of loneliness.

For a brief second, he felt the weight of sadness pressing on him.

But instead of pushing it away, he allowed himself to feel it fully.

And as he did, the gray orb began to fade, blending into the other colors—no longer overwhelming, but simply a part of the whole.

When Paolo opened his eyes, he felt more at peace than he had in years.

He looked at the data on the screen—the graph showed a rare equilibrium between gamma, alpha, and beta waves.

A sign that he had reached emotional balance.

He wrote in his journal:

"Emotions are not enemies to fight but parts of us to embrace. Only by accepting them can we find balance."

Chapter 24:

Francesca - The Garden of Desires

✦ ✦ ✦

Francesca woke up feeling **light**, as if the **night had carried away a weight she didn't realize she was carrying**.

The **morning sun** filtered through the curtains, filling the room with **a golden glow**.

She prepared a **cup of tea** and stepped onto the **balcony**, letting the **warmth of the sun touch her face**.

That morning, she felt a **pull toward something new**—a desire to **explore the deepest corners of her mind**.

She placed the **eMind EEG** device on her head and **closed her eyes**.

Her mind took her to a **magical garden**, the **most beautiful she had ever seen**.

The **flowers were impossible colors**—deep purples, glowing turquoise, golden petals that shimmered like sunlight.

The **air was filled with a perfect mix of lavender, citrus, and sweetness**—a fragrance that wrapped around her like a warm embrace.

She walked along **narrow stone pathways**, her **bare feet brushing against the soft grass**.

At the **center of the garden**, there was a **fountain**, but instead of water, it flowed with **golden light**, small **sparkles of energy rising into the air like fireflies**.

Francesca approached, **fascinated**.

Each **golden spark** contained **an image**—a **dream, a wish, a hidden desire**.

It was as if the **fountain held the essence of everything she had ever longed for**.

She reached out and **touched one of the golden lights**—and suddenly, she was transported to **a vision of her future**.

She stood in a **bright, open home**, surrounded by **books, paintings, and warmth**.

The windows let in **golden afternoon light**, and outside, a **field of sunflowers stretched endlessly**.

There was **laughter in the distance**, and as she turned, she saw **people she loved**, their faces glowing with happiness.

She wasn't just **imagining** this scene.

She was **feeling it—living it**.

Then, she touched another **golden spark**, and a new **vision unfolded**:

She was standing **by the ocean at sunset**, feeling **the waves lap at her feet**, the sky painted in hues of pink and gold.

A **deep peace filled her**, a sense of **belonging to something greater**.

For the first time, she understood that **her desires were not just fantasies—they were possibilities**.

They were **pieces of herself, waiting to be realized**.

When she opened her eyes, she felt **transformed**.

She removed the **device** and looked at the **data on the screen**—the **gamma and alpha waves were in perfect harmony**, showing that her mind had reached a **deep state of emotional clarity**.

She wrote in her journal:

"Desires are like seeds. Plant them in your mind, nurture them with love, and they will grow into reality."

She closed her journal with a **smile**, knowing that her **imaginary garden** had just shown her the **path to everything she truly wanted**.

Chapter 25:

Paolo - The Conscious Choice

✦ ✦ ✦

Paolo paced back and forth in his study, the **eMind EEG** headband in his hands.

He kept looking between the device and the window, as if waiting for a sign to begin.

For days, he had felt that something important was about to happen, and for the first time in a long while, he knew he couldn't ignore it.

He sat at his desk, placed the headband on his forehead, and started the app.

Immediately, his mind created an image:

A steep, winding road, cutting through rugged hills.

Paolo recognized it instantly—a symbol of his current life.

A clear, defined path, but one that was narrow, structured, predictable.

Each turn in the road represented a goal he had reached, a success he had achieved.

But as he walked, he felt a strange emptiness.

Where was this path leading him?

He couldn't see the destination, and for the first time, the certainty of the road ahead felt meaningless.

The scene shifted, and suddenly he found himself in a vast open field.

Golden light bathed the endless landscape, stretching toward the horizon with no barriers, no roads, no limitations.

There were no boundaries, no predetermined routes—just pure, infinite space.

Paolo stepped into the field, feeling the grass brush against his fingertips, the warm sun on his face.

For the first time in years, he felt free.

He ran—not toward a goal, not toward an expectation, but simply because he could.

And as he ran, he laughed, feeling a joy he hadn't known he was missing.

Then, the scene shifted once again.

He found himself in a dark room, standing before an ancient wooden door.

The door was solid, heavy, and completely closed.

For a moment, he hesitated.

There was no way of knowing what was beyond it.

But something inside him whispered:

"This is the choice that truly matters."

He took a deep breath, reached out, and opened the door.

On the other side, there was no physical place—just a mirror.

His own reflection stared back at him, but this time, something was different.

His reflection wasn't distant, or empty, or uncertain.

It was calm, knowing, understanding.

And suddenly, he realized the truth:

It didn't matter which road he took.

The path itself was never the answer.

What mattered was how he chose to walk it.

When he opened his eyes, his breathing was steady, his mind clear.

The brainwave graph showed high spikes in both gamma and beta waves, a sign that he had processed something deeply significant.

He removed the device and leaned back in his chair, feeling a peace he had never experienced before.

He wrote in his journal:

"The most important choice is not between two roads—it is between who we are and who we want to become."

Chapter 26:

Francesca - The Inner Light

✦ ✦ ✦

Francesca sat on her **balcony**, wrapped in a **light scarf**, as the **sunset painted the sky in orange and violet hues.**

The crisp evening air carried the **scent of autumn**, filling her with a sense of **renewal.**

That night, she felt the need to **connect with a part of herself that she often forgot**—the **inner light that made her feel alive, authentic, and full of hope.**

Taking a **deep breath**, she placed the **eMind EEG** device on her head and closed her eyes.

She let her **mind drift**, and soon, she found herself on a **hill she knew well.**

It was a place from her **childhood**, a **sanctuary from the past**, where she often came to **escape and dream.**

The **endless green field stretched before her**, dotted with **small white flowers swaying in the breeze.**

In the distance stood a **solitary tree**, its golden leaves shimmering in the light of the **setting sun**.

Francesca began to **walk toward it**, feeling the **soft grass under her feet**, the **wind's gentle caress on her skin**.

When she reached the tree, she **sat down beneath its branches**, resting her back against its **sturdy trunk**.

She closed her eyes, letting the **warm light of the sun filter through the leaves**, and listened to the **whispering wind** that seemed to **speak in an ancient language**.

Then, something **unexpected happened**.

A **soft glow** appeared before her.

It wasn't **blinding**, but **gentle and warm**, like a presence **watching over her**.

At first, she thought it was **something external**.

But then she **realized the truth**—

The **light wasn't coming from outside**.

It was coming from **within her**.

A **manifestation of her energy, her essence**—the part of her that had **always been there**, waiting to be **seen, acknowledged, and embraced**.

Slowly, the **light expanded**, reaching out like **a warm embrace**.

Francesca extended her hand and **touched it**, and in that moment, she felt **pure warmth spreading through her chest**—a sensation of **completeness**.

She had **found herself again**.

When she **opened her eyes**, she was still **on her balcony**, the night sky now filled with **countless stars**.

She removed the **device** and glanced at the **brainwave graph**—her **gamma and alpha waves were at their peak**, a sign that her **mind had reached a deep emotional balance**.

She picked up her journal and wrote:

"The light we seek outside is already within us. It is not hidden—we are just too distracted to see it. Sometimes, all we need to do is stop, breathe, and listen."

Chapter 27:

Paolo - Overcoming Doubt

✦ ✦ ✦

Paolo sat at his **desk**, staring at the **eMind EEG** device with a **contemplative expression.**

It had become his **evening companion**, a bridge between the **logic that governed his life** and the **world of emotions and imagination that he was only beginning to understand.**

But that night, he **couldn't bring himself to start**.

A doubt—**persistent and stubborn**—kept creeping into his thoughts.

Running a hand through his hair, he studied the device as if it were **challenging him.**

"What if all of this is just an illusion?"

Despite the **data, the experiences, the emotions he had felt**, a part of him still **resisted.**

Was his **mind truly capable of such experiences**, or was it simply **tricking itself**?

Taking a deep breath, he placed the **headband on his forehead**, started the app, and closed his eyes.

"Tonight, I won't avoid the doubt. I will face it."

His mind **constructed an image**—

A **long, dark corridor**, lined with **mirrors on both sides**.

But when he **looked into them**, his **reflections were blurred**, indistinct.

As he walked forward, the mirrors **shifted**, warping his image.

"Why can't I see myself clearly?" he whispered.

The corridor seemed to **stretch endlessly**, every step filling him with **frustration and confusion**.

Then, **a soft glow appeared ahead**.

A **figure emerged from one of the mirrors**—

A **younger version of himself**, with **eyes full of hope and a hint of uncertainty**.

For a moment, they **stood face to face**.

The younger **Paolo didn't speak**—but his **presence communicated everything**.

A feeling of **understanding, of reassurance**.

"You don't need all the answers," the figure seemed to say.

"Doubt is not failure—it's growth."

A warmth spread through **Paolo's chest**.

And as he **accepted the uncertainty**, the **mirrors stopped shifting**.

His **reflection became clear again**.

When he **opened his eyes**, he saw that his **gamma and beta waves had spiked**, indicating a **deep cognitive and emotional realization**.

He removed the device and wrote in his journal:

"Doubt is not the enemy—it is a door. The fear of stepping through it is what holds us back."

Chapter 28:

Francesca - The Courage to Let Go

✦ ✦ ✦

Francesca sat on the **sofa**, her **journal open on her lap**.

She slowly flipped through its pages, her fingers tracing **the hurried words, the rushed thoughts, the fragments of emotions captured in ink.**

Yet one theme **kept surfacing**.

The **past still held her tightly**.

She took a **deep breath** and placed the **eMind EEG** device on her forehead.

This time, she wasn't **searching for inspiration.**

She was going to **face the one thing she had avoided for too long**:

Letting go.

Her mind took her to a **bright, white room.**

A **door stood before her**, waiting to be opened.

She hesitated.

Then, gathering **all her strength**, she **pushed the door open**.

Beyond it lay a **vast meadow**, golden grass swaying under a warm, gentle breeze.

At the **center of the field**, stood **an ancient tree**.

At its roots, there was a **small wooden box**.

As she approached, her **heart pounded**.

She knew what was inside.

She **knelt** and lifted the **lid**.

Inside, she saw **memories of pain, of things she had tried to forget**.

Moments of **loss**, of **loneliness**, of **words left unsaid**.

Her chest tightened, but she **did not look away**.

Instead, she **gently closed the box** and dug a **hole beneath the tree**.

She placed the box inside and **covered it with earth**, her hands trembling.

As she **stood up**, she looked at the tree—it **seemed stronger now**, its leaves **shining in the golden light**.

It was as if her **pain had transformed**—not **disappeared**, but **become part of something greater**.

When she **opened her eyes**, she felt **lighter than she had in years**.

The **brainwave graph showed deep emotional processing**, with **waves of both release and resolution**.

She removed the device and wrote in her journal:

"Letting go is not losing. It is making space for something new—something we deserve to welcome."

Chapter 29:

Paolo - Rediscovering Himself

✦ ✦ ✦

Paolo ran a hand over his face, staring at the **eMind EEG** device resting on the table.

It felt almost like a magical object, capable of unveiling truths he had long forgotten.

For days, an image had kept resurfacing in his mind—

A reflection of himself, younger, more hopeful, filled with energy.

He couldn't shake off the feeling that something had been left behind—something essential.

Finally, he sat at his desk, placed the headband on his forehead, and closed his eyes.

His mind created an image—

A high school science lab, filled with the buzz of students and the scent of books and chemicals.

He saw his younger self, surrounded by experiments and circuits, his eyes wide with fascination.

The room was full of excitement, a place where everything seemed possible.

The sounds of laughter, the thrill of discovery, the simple joy of curiosity.

Paolo watched as the teenage version of himself eagerly wrote formulas on the board, explaining theories to his classmates.

"It was so simple back then," he thought.

There were no limits, no doubts—just pure passion for learning and discovery.

The scene shifted.

Now, he was in his present-day office, surrounded by monitors, graphs, and neatly stacked papers.

The glow of the screen illuminated his face, showing a man who had mastered his field but had lost something along the way.

The passion had been replaced by precision.

The curiosity had been replaced by responsibility.

Then, something unexpected happened.

The two versions of Paolo sat across from each other—

The teenager, filled with wonder and ambition,

And the man, carrying years of experience and success.

For a moment, they simply observed each other.

The young Paolo examined the projects on the desk, nodding in admiration.

The older Paolo listened to the excitement in the younger one's voice, remembering the joy of learning without limits.

Then, as if they both understood the same truth,

They smiled at each other.

"You don't have to choose between past and present," his younger self seemed to say.

"You can be both."

When Paolo opened his eyes, his chest felt lighter.

He removed the device and looked at the data—his gamma waves were at an all-time high, signaling a powerful realization.

He picked up his journal and wrote:

"We are not forced to choose between who we were and who we are. We can integrate both to build the future we truly want."

Chapter 30:

Francesca - Finding Lost Time

✦ ✦ ✦

Francesca sat by the **window**, listening to the **gentle rhythm of rain** against the glass.

She prepared a **cup of tea**, letting the **steam warm her hands**.

That evening, she felt the **weight of time**—

Not **the time that moves forward**, but the **time we lose**, the **moments we don't live fully**.

She placed the **eMind EEG** on her forehead and closed her eyes.

The first image that appeared was **a memory from childhood**.

She was in **her mother's kitchen**, helping to **knead dough**, the **scent of fresh bread filling the air**.

Her **small hands followed her mother's movements**, while **laughter and warmth surrounded her**.

She could feel the **flour dusting her fingertips**, the **sound of her mother's gentle encouragement**.

A moment **so simple, yet so full of love.**

Then, the memory **shifted**—

She was now **a teenager**, sitting on a **stone wall**, a **notebook open on her knees.**

The **sunset painted the sky in warm shades**, and she was **writing poetry**, her mind **overflowing with words and emotions.**

She even remembered one of the lines she had written:

"Life is the wind, and we are leaves dancing within it."

At the time, she had thought it was **a childish idea.**

But now, those words seemed to **hold deep wisdom.**

More images **came to her**—

A **summer night spent stargazing with a friend,**
A **walk in the rain that had turned a dull afternoon into magic,**
A **moment of pure silence where everything felt whole.**

Fragments of her past, coming back to **remind her**—

The **most beautiful moments in life** are often **the simplest ones.**

When she **opened her eyes**, her **breath felt slower, steadier.**

She glanced at the **graph on the screen**—it showed **perfect balance between relaxation and emotional engagement.**

She removed the **device** and wrote in her journal:

"Time is never truly lost. It stays within us, hidden in our memories, waiting to be rediscovered when we have the courage to look back."

Chapter 31:

Paolo - The Strength of Forgiveness

✦ ✦ ✦

Paolo sat in front of his **desk**, the **eMind EEG** headband in his hands.

Tonight, he knew there were **no more excuses.**

He had postponed this moment **for too long**—hiding behind his **routine, his rational mind, his work.**

But the **weight of an old wound had become unbearable.**

He placed the **device on his forehead**, took a deep breath, and **closed his eyes.**

His mind took him to a **familiar place**—

A **quiet park**, bathed in the golden light of **late afternoon.**

There, sitting on a **wooden bench**, was **a figure from his past.**

An **old friend**—someone with whom he had **shared years of laughter and trust,**

Until a **fallout had severed their bond.**

Paolo **froze**, his chest tightening.

The pain of **unspoken words, unresolved anger, and the years of silence** weighed heavily on him.

But his **friend**—or rather, the **memory of him**—simply **smiled.**

There was **no resentment in his expression**, only **understanding.**

Paolo hesitated, then **took a deep breath** and walked over.

He sat down beside him.

No words were needed.

The **silence between them wasn't empty**, but **filled with meaning.**

Then, without thinking, Paolo **extended his hand.**

His friend **took it**, and in that moment, Paolo felt **something break open inside him**—

Not in **pain**, but in **relief**.

He realized that **forgiveness was not about the other person**—

It was about **freeing himself from the burden of carrying the past**.

When he **opened his eyes**, a **wave of peace settled over him.**

He removed the **device** and looked at the **data**—his **gamma waves had surged**, a sign of **deep emotional processing and release.**

He wrote in his journal:

"Forgiveness is not a sign of weakness. It is a gift we give ourselves—to live unchained from the past, ready to move forward."

Chapter 32:

Francesca - The Roots of the Heart

✦ ✦ ✦

Francesca sat under the starry sky, leaning against the trunk of an old tree in the park near her home.

She had chosen this quiet corner of nature as her personal sanctuary—a place where she could think freely.

That evening, with the cool breeze whispering through the leaves, she felt the need to explore the roots of her heart—to understand where she had come from and where she was going.

She placed the **eMind EEG** device on her forehead and closed her eyes.

Her mind took her to a familiar garden, a place she had visited many times in her imagination.

But this time, it felt more vivid—more real than ever before.

At the center of the garden stood a majestic tree, its roots stretching deep into the earth, its branches reaching toward the stars.

Francesca knelt beside it, running her fingers over its rough bark.

The moment she touched the roots, a gentle vibration spread through her fingertips.

And suddenly—memories flooded in.

Each root she touched revealed a different moment in her life:

Sitting on her grandmother's lap, listening to stories of a world she had never known

Laughing with her best friend under a summer sky, feeling as though time would never separate them

Writing secret dreams in her childhood diary, never daring to say them out loud

The memories were warm, comforting—reminders of where she came from, the people who had shaped her, the love that had nourished her.

Then, she touched a thicker, older root, and a different memory emerged—

A moment of loss, of uncertainty.

She saw herself sitting alone in her room, feeling disconnected.

She had forgotten these moments—or perhaps, she had chosen to ignore them.

But as she sat beside the tree, she realized something important—

Every root mattered.

The joyful memories gave her strength.

The painful ones taught her resilience.

They were all part of her.

When Francesca opened her eyes, she felt a deep sense of connection—not just with her past, but with herself.

She removed the device and looked at the data—her gamma and alpha waves were perfectly balanced, a sign that her mind had reached a profound emotional clarity.

She wrote in her journal:

"Our roots define us—not by keeping us stuck, but by showing us how far we can grow."

Chapter 33:

Paolo - The Silence That Speaks

✦ ✦ ✦

Paolo sat in his dimly lit study, the ticking of the clock the only sound in the room.

For a man who had always lived in structured schedules, constant noise, and fast decisions, silence had never been a companion—only an absence of sound.

But that night, he wanted to experience it differently.

He placed the **eMind EEG** on his forehead and closed his eyes.

For the first time, he didn't try to imagine anything.

He simply waited.

At first, there was nothing but darkness.

Then, slowly, the silence began to take shape.

He could hear his own breathing, the steady rhythm of his heartbeat.

He became aware of the subtle tension in his shoulders, the way his fingers twitched slightly when he was deep in thought.

It was as if his body had been speaking to him all along, but he had never truly listened.

And then, without warning, memories began to surface.

Moments he hadn't thought about in years—

Sitting on the beach as a child, watching the waves crash against the shore

His father's voice, teaching him how to fix a broken bicycle chain

The quiet satisfaction of finishing a book he loved, long after the world had fallen asleep

The silence wasn't empty—

It was full of life, full of memories, full of everything he had overlooked.

When he opened his eyes, the graph showed something surprising—

His alpha waves were dominant, meaning his mind had entered a state of deep peace and awareness.

He removed the device and wrote in his journal:

"Silence is not the absence of sound. It is the presence of everything we have ignored."

Chapter 34:

Francesca - The Ocean of Thoughts

✦ ✦ ✦

Francesca sat cross-legged on the carpet, holding a cup of tea that had long gone cold.

Her mind was a whirlwind of thoughts, emotions pulling her in a hundred different directions.

She needed to let go, to find stillness in the storm.

She placed the **eMind EEG** on her forehead and closed her eyes.

The moment she did, her mind created an image—

A vast ocean, stretching endlessly to the horizon.

The water was calm at first, waves rolling in softly, rhythmically.

She walked along the shore, feeling the warm sand between her toes.

With each step, a thought surfaced—

A conversation she never finished

A decision she had put off

A dream she had buried deep inside

The waves carried her worries to shore, but instead of fighting them, she simply observed.

She stepped into the water, feeling it cool against her skin.

And as she waded deeper, she let the waves carry away her doubts.

Until all that remained was peace.

When she opened her eyes, her mind felt clearer than it had in weeks.

She removed the device and wrote in her journal:

"The ocean does not hold onto its waves—it lets them rise and fall. So should we with our thoughts."

Chapter 35:

Paolo - The Bridge Between

Past and Future

✦ ✦ ✦

Paolo's mind was full of **questions.**

How do we **move forward** while still **honoring where we came from**?

He placed the **eMind EEG** on his forehead and closed his eyes.

Immediately, an image **appeared**—

A **long bridge**, stretching between two towering cliffs.

Behind him—**his past**, filled with **memories, choices, regrets, victories.**

Ahead—**a future shrouded in mist**, unknown, uncertain.

The bridge **wasn't steady.**

It **swayed in the wind**, challenging him **to trust the path ahead.**

As he **walked forward**, he looked back—

And he saw that **his past was not disappearing.**

It was **becoming part of the bridge itself**.

"The past is not meant to be left behind," he realized.

"It is meant to carry us forward."

When he opened his eyes, he felt **lighter**.

He wrote in his journal:

"The past is not a weight—it is the foundation of our future."

Chapter 36:

Francesca - The Power of Gratitudez

✦ ✦ ✦

Francesca closed her journal with a smile, letting it rest on the table beside a steaming cup of tea.

It had been a day full of reflection, and tonight, she felt the need to dedicate a moment to something she often overlooked: gratitude.

The past few weeks had transformed her way of seeing the world, and she knew that the time had come to give thanks—not just to others, but to herself as well.

With calmness, she put on the **eMind EEG** and allowed herself to let go, closing her eyes.

The first image that emerged was that of a small, cozy room, with cream-colored walls and a large chalkboard covering an entire wall.

On the table in front of her lay a piece of white chalk.

Francesca picked it up, feeling a strange emotion rise within her—a mix of excitement and sweet nostalgia.

She approached the chalkboard and began to write.

The first word was "family."

Immediately, images of moments spent with her mother, father, and sister resurfaced.

She saw their smiling faces during holiday dinners, the laughter they shared, and even the arguments that, in the end, had only strengthened their bond.

Then, she wrote "friends."

Each letter seemed to evoke a different memory:

Conversations under the stars

Evenings spent sharing secrets

Acts of kindness during difficult times

Every image filled her heart with warmth.

As she continued writing other words—

"health," "opportunities," "nature," "love"—

the chalkboard became covered in bright, glowing words.

Each one was a testimony to the gifts life had given her, often without her even realizing it.

The imaginary room began to light up, as if those words were beacons dispersing any lingering shadows of dissatisfaction.

Francesca stood for a long moment, gazing at the chalkboard.

Each word vibrated with a special energy, a reminder of abundance and beauty.

She no longer felt lacking—she felt full, overflowing with gratitude for everything she had experienced.

When she opened her eyes, the graph on the screen showed a perfect balance between alpha and gamma waves.

She removed the headband and wrote in her journal, with a newfound clarity and joy:

"Gratitude is not just an emotion. It is a lens through which we see the world. It is the key to transforming what we have into everything we need."

Chapter 37:

Paolo - The Encounter with the Unknown

✦ ✦ ✦

Paolo sat at the **kitchen table**, the **device in front of him** and a **cup of coffee, now cold, by his side.**

The past few weeks had **taught him a lot about himself**, but tonight, he felt something **different.**

The **unknown was calling him**—not as a **threat**, but as a **possibility**.

He wondered what it truly meant to **make space for what he did not yet know.**

Paolo had always been **a man of certainties**, an **architect of his own life**, building everything with **precision and order.**

The **unknown** had never been **part of his plans.**

But now, with **eMind EEG**, he had found a way to **explore those uncharted territories**—not with **fear**, but with **curiosity**.

He put on the **headband** and closed his **eyes.**

He found himself in a **landscape shrouded in thick fog.**

The **ground beneath his feet was soft**, like **damp grass**, and the air was **cool, almost sharp.**

Each **step forward** was accompanied by a **gentle whisper**, as if **the unknown itself were inviting him to proceed.**

There were **no roads, no signs**—only the **feeling that something was waiting beyond the fog.**

After a while, a **large wooden door emerged from the mist.**

It was **massive and ancient**, framed by **intertwined branches**, as if it had grown **out of the earth itself.**

The **wood grain seemed to pulse**, as though the door was **alive, a part of something greater.**

Paolo hesitated, then **reached out and pushed.**

The **door opened slowly**, revealing a **circular room**, bathed in **soft, golden light.**

The **walls were covered in mirrors**, each reflecting a **different image.**

He stepped forward and **approached the first mirror.**

In it, he saw **his younger self**—his **eyes filled with ambition and hope.**

In the **second mirror**, he saw **himself as he was now**—his face **marked by the years**, yet carrying the **wisdom that only experience can bring.**

Each mirror showed a **fragment of who he had been or who he could have become.**

At the **center of the room**, a **glowing, shifting light hovered**, flickering like a **living flame.**

With every **breath he took**, the light **changed color**—from **blue to red, to green, to white**.

Paolo felt his **heartbeat quicken** as he stepped closer.

The **light responded**, pulsing in rhythm with his **own heartbeat**.

"What are you?" he asked.

No **words** came in response—**only a feeling**.

Not **fear**, but **possibility**.

Not **limitation**, but **openness**.

In that moment, **Paolo understood**.

The **light represented everything he had yet to discover—the unknown parts of himself and the world**.

The **things he had always avoided, dismissed, or left unexplored**.

Slowly, he **extended his hand** and **touched the light**.

The **mirrors dissolved**, and in their place came **a vast sense of peace**—

A realization that **the unknown was never something to fear, but something waiting to be embraced**.

Paolo **opened his eyes**, finding himself **back in his kitchen**.

He looked at the **graph on the screen**—his **gamma waves were at their peak**, a sign of **intense cognitive activity and deep realization**.

He removed the **headband** and stared at the **empty space in front of him**, reflecting on what he had just experienced.

Then, he picked up his **journal** and wrote:

"The unknown is not an enemy. It is a part of us waiting to be discovered. It is where change and potential are born."

Chapter 38:

Francesca - The Awakening of Trust

✦ ✦ ✦

Francesca wrapped herself in **her favorite blanket** and sat by the **open window.**

The **cool evening air** carried the scent of **fresh rain**, a fragrance that always **comforted her.**

But that night, she needed **something more**—

She needed to **rediscover trust.**

Not just **in others**, but **in herself.**

She put on the **eMind EEG** and closed her **eyes**, allowing the app to **record every thought.**

At first, her mind was **empty**, like a **blank page.**

Then, slowly, an **image emerged**—

A **forest bathed in the golden light of morning.**

The **trees stood tall and majestic**, their **branches weaving a canopy of leaves overhead.**

The **crunching of dry leaves beneath her feet**, combined with the **gentle song of birds**, created a **peaceful, reassuring melody**.

Francesca **walked slowly**, following a **natural path** that led to the **heart of the forest**.

There, in a **small clearing**, stood an **ancient well**, surrounded by **moss and wildflowers**.

She **stepped closer**, curiosity pulling her toward it.

She peered into the **clear water**, and there, she saw **her reflection**.

It was **her**, but not quite.

The image was **clear, yet imperfect**—

She could see the **traces of her insecurities**, the **faint outlines of her fears**.

She gazed at her **reflection for a long moment**, then **closed her eyes**—

Even in her **imagination**, she allowed herself a **pause**.

And when she **opened them again**, the **reflection had changed**.

Now, she saw a **stronger version of herself**—

Her eyes **full of determination**,
Her lips curved into **a confident smile**,
The expression of **someone who had learned to trust in her own strength**.

Then, a **gentle but firm voice** whispered within her—

"You have everything you need. Trust yourself. You can do this."

That feeling **wrapped around her** like a **warm embrace**.

She **stepped away from the well** and returned to the **path**,

But this time, the **forest felt different**—

Brighter, more **alive**.

Every **tree seemed to whisper encouragement**, Every **ray of sunlight felt like a guiding hand**.

She **opened her eyes**, finding herself **back in her living room**.

The **graph on the screen** showed **perfect harmony between alpha and gamma waves**—

A sign of **inner balance and emotional clarity**.

She removed the **headband** and sat **in silence** for a few minutes,

Feeling a **sense of serenity she hadn't felt in a long time.**

She picked up her **journal** and wrote:

"Trust is not the absence of doubt. It is the strength to move forward, knowing that within us, we have everything we need to face any challenge."

Chapter 39:

Paolo - The Strength of Forgiveness

✦ ✦ ✦

Paolo sat at his **desk**, holding the **eMind EEG** device in his hands.

Tonight, he knew there were **no more excuses.**

For too long, he had **postponed this moment**, hiding behind his **rational mind and daily routine.**

But the **weight of what he carried inside him had become unbearable.**

He had to **face it.**

He had to **let it go.**

And forgiveness—**as difficult as it was**—was the **only way forward.**

He placed the **headband on his forehead** and closed his **eyes**, letting his mind **guide him where it needed to go.**

He found himself on a **path bathed in golden light.**

The **trees on either side whispered soft words of comfort,**

While a **gentle breeze** caressed his face.

Paolo walked **slowly**, knowing that **at the end of the path, he would find what he had been avoiding for years.**

The path opened into a **peaceful clearing**,

Where a **large tree stood at the center.**

Beneath it sat a **wooden bench**,

And on that bench was a **figure Paolo knew well**—

A **friend from his past**,

Someone with whom he had **shared years of laughter and trust**,

But whom he had **lost after an argument that had shattered everything.**

The friend looked at him **with a serene smile**,

As if time had **never erased what had once united them.**

Paolo **hesitated**, his heart pounding.

A mix of **fear, regret, and longing for reconciliation swirled within him.**

Finally, he found the **courage to step forward.**

He sat down **beside his friend.**

No words were needed.

The **silence between them was not empty**, but **full of meaning.**

Paolo felt the **weight of old accusations, misunderstandings, and words left unspoken.**

Then, with a **simple but powerful gesture**, he **extended his hand**.

His friend took it.

And in that moment, Paolo felt **a wave of warmth rush through him**.

There were no **justifications, no elaborate apologies.**

Only **acceptance**.

Only **the willingness to let go of the pain that had kept them apart**.

And then, he **understood**—

Forgiveness was not for the other person.

It was for **himself**.

To free himself from a **burden that had held him back for too long**.

When he **opened his eyes**, he felt **lighter than he had in years**.

He glanced at the **graph on the screen**—

His **gamma waves had peaked**, a sign that his mind had **processed deep emotions**.

He removed the **device**, reached for his **journal**, and wrote:

"Forgiveness is not weakness, but strength. It is the gift we give ourselves—to live free from the past and ready to look ahead."

Chapter 40:

Francesca - The Gift of Empathy

✦ ✦ ✦

Francesca sat on the living room floor, her journal open before her, the warm glow of a candle flickering in the dimly lit room.

That evening, she felt the need to go beyond herself—to connect with the emotions of others.

Ever since she had begun her journey through imagination, she had discovered how important it was to see the world not only through her own eyes but also through the eyes of those around her.

She placed the **eMind EEG** on her forehead and closed her eyes.

Immediately, she found herself in a white room, bathed in soft, soothing light.

At the center of the room, there was a mirror.

But when she stepped closer, she realized that it did not reflect her image.

Instead, the mirror began to change, revealing unknown faces and scenes from everyday life.

Each image carried an emotion, and Francesca found herself immersed in each one.

The first image showed an elderly woman sitting alone on a park bench, her eyes lost in the emptiness of solitude.

A wave of loneliness washed over Francesca, a deep sense of abandonment tightening in her chest.

She imagined herself sitting beside the woman, gently taking her hand.

The woman's face lit up with a smile, and Francesca felt a deep connection,

As if that small gesture had filled a void that had existed for years.

The scene shifted, revealing a child playing alone in a messy room.

His eyes were fixed on the window, filled with hope, but also tinged with sadness.

Francesca knelt beside him, imagining herself offering comfort and reassurance.

The child turned toward her, and as he smiled, the room filled with a warm, golden light.

Every image in the mirror told a different story—

A man overwhelmed by work, lost in endless tasks

A young woman struggling with her self-image, searching for self-acceptance

An elderly man walking slowly down the street, carrying the weight of years of solitude

Francesca didn't just observe these moments.

She lived them.

She felt the pain, the joy, the hope, and the fear—as if they were her own.

And in that moment, she understood something profound—

Empathy is the bridge that connects us to one another.

Through it, we can see that we are never truly alone.

Because every heart carries a story worth listening to.

When Francesca opened her eyes, she felt changed.

The graph on the screen showed a surge in gamma waves,

A sign that her mind had processed deep, complex emotions.

She removed the headband, picked up her journal, and wrote:

"Empathy is the bridge that connects us to others. Through it, we see that we are never truly alone, because every heart carries a story worth listening to."

Chapter 41:

Paolo - The Strength of Vulnerability

✦ ✦ ✦

Paolo sat motionless at his **desk**, his fingers lightly tracing the smooth surface of the **eMind EEG**.

Tonight felt **different from the others**—one of those nights when the **silence around him seemed to amplify the noise within him**.

Vulnerability.

A word he had always **avoided**, a **territory he considered dangerous and unstable**.

And yet, he knew there was **something powerful within it**, something that kept **calling to him**.

Taking a **deep breath**, he placed the **headband on his forehead** and closed his **eyes**.

His mind immediately took him to **a familiar place**—his **old office**.

The **image was sharp, almost tangible**.

The **hum of the computer**, the **rhythmic tapping of the keyboard**, the **artificial light casting a sterile glow over the room.**

A scene he had **lived through countless times**—

The feeling of **carrying everything on his shoulders,** Of needing to **prove something to everyone**—except **himself.**

Paolo **watched the scene as an observer,**

Seeing his **past self hunched over the desk**, his hands **holding his head in exhaustion.**

He could **feel the weight of those moments**, the **pressure of expectations pressing down on him.**

And for the **first time**, he allowed himself to **feel compassion for himself.**

There was **no judgment,**

Only a **sincere acknowledgment of how difficult those moments had been.**

The **scene dissolved**, and Paolo found himself in **an empty room.**

The **floor was made of pale wood**, and at the **center of the room stood an old wooden chest,**

Its surface **worn by time.**

His hands **trembled slightly** as he approached and lifted the **lid.**

Inside were **fragments of his past**—

Objects that **represented moments of vulnerability he had tried to forget.**

A **letter**, yellowed with age, carefully folded.

He picked it up, remembering the **words he had written years ago to someone important**,

Words he had never found the **courage to send.**

He felt the **weight of that moment**,

But also the **liberation of holding it in his hands.**

That letter was **no longer a symbol of failure**—

It was a **reminder that he had once tried to be honest**,

Even if that honesty had remained **unfinished.**

Next to the **letter**, he found an **old notebook with missing pages.**

He flipped through it, recognizing his **own messy handwriting and scribbled notes.**

They were **thoughts from a difficult time**,

Words expressing a **deep longing to be understood, to find direction.**

What had once made him **ashamed** now felt like a **testament to how far he had come.**

Finally, he found a **small photograph.**

It was from a **day he had always tried to erase from his memory.**

As he **looked at it**, a **tightness gripped his chest.**

But instead of **looking away**, he forced himself to **really see it.**

He **relived the pain of that moment**,

But along with it came the **realization of how that pain had shaped him**,

How it had contributed to **making him the man he was today**.

As if sensing it had **fulfilled its purpose**,

The **chest slowly closed on its own**.

Paolo then noticed a **glass wall before him**.

His **reflection stared back at him**.

It was an **imperfect image**—

Lines on his face **telling stories of struggles**,

His **eyes carrying traces of difficult moments**.

But in that **imperfection**, he saw **something real**.

He saw **strength**.

He saw **humanity**.

When he **opened his eyes**, he felt a **lightness he hadn't felt in years**.

He glanced at the **graph on the screen**—his **gamma waves showed intense activity**,

A sign that his mind had **processed deep emotions**.

He removed the **device** and took a moment to **savor the feeling**.

It wasn't a **victory**,

But it was **a silent, intimate breakthrough**.

He picked up his **journal** and wrote:

"Vulnerability is not a weakness to hide, but a door that leads us to who we truly are. In our fragility, we discover our greatest strength."

Chapter 42:

Francesca - The Strength of Connection

✦ ✦ ✦

Francesca stood by the **window for a long time**, gazing at the **stars**.

Each **shining point in the sky** seemed to **tell a story**, a **fragment of life intertwining with others to form a greater picture**.

But that night, she felt a **deep sense of isolation**, one she had tried to **ignore for too long**.

It was as if her **journey inward** had created a **distance from others**,

A distance she now **longed to bridge**.

She placed the **eMind EEG** on her forehead,

Allowing the **app to record every thought and emotion**.

As she closed her **eyes**, she found herself in an **enchanted garden**,

With **pathways branching out in every direction**.

The **air was crisp and fragrant**, and the **rustling of the leaves felt like an invitation.**

At the **center of the garden stood a large circular table**, completely **empty**.

Francesca **stepped closer** and noticed that **small pathways led away from the table**,

Each one connected to a **smaller table in the distance.**

Her **heart beat slightly faster** as she took her **first steps down one of the paths.**

At the end of the trail stood a **table adorned with an old pocket watch**,

Its surface **shiny and well-preserved.**

She **sat down and gently touched the watch.**

Immediately, a **familiar scene came to life**—

Her **grandfather**, seated in his **armchair**,

Telling a **story with his warm, reassuring voice.**

She remembered **how he would always remind her**

That **time was precious**,

But that she should never be **afraid to slow down**

And truly **appreciate the important moments in life.**

A **deep gratitude** filled her chest,

As those words now **resonated within her more than ever before.**

She continued down **another path**,

Which led to a **table with an open notebook.**

The **paper was worn**, as if it had been **flipped through thousands of times.**

She recognized **the small symbol her best friend used to draw.**

She **relived the carefree laughter of a summer afternoon,**

The **shared dreams of sleepless nights,**

But also the **silences that had grown between them over the years.**

She realized that **despite the distance,**

That **friendship was still a part of her.**

She **decided, right then and there,**

That she would **find a way to rebuild that bond**—

Maybe with **a phone call,**

Or even **a simple but heartfelt message.**

When she reached the **third table,**

She saw **a chipped coffee cup** resting on its surface.

She **hesitated for a moment,**

Knowing exactly **what it represented.**

But instead of **turning away**, she **sat down and studied it.**

She **relived a past relationship,**

Remembering the **beautiful moments,**

But also the **difficult ones.**

There was no longer **anger**—

Only a **calm understanding** of what had been.

Francesca imagined **a conversation she had never had,**

One where she allowed herself to **forgive both herself and the other person.**

Accepting that **some stories end,**

But that they **always leave behind something valuable.**

Finally, she reached the **empty table.**

It seemed **simple**, yet she could sense **its infinite potential.**

A feeling of **excitement and hope** filled her.

This table symbolized **the connections she had yet to make,**

The **people she had yet to meet,**

The **experiences waiting for her.**

She imagined herself **smiling at unfamiliar faces,**

Creating **new bonds based on honesty and trust.**

When she **opened her eyes,**

The **garden faded**, but the **sense of connection remained.**

The **graph on the screen showed perfect balance between gamma and alpha waves,**

A sign that she had reached **a state of deep awareness and serenity.**

She removed the **device** and sat in silence for a moment,

Savoring the **profound peace she felt.**

With **steady hands**, she picked up her **journal** and wrote:

"Connection is the bridge that links who we are to who we can become. It is the thread that ties us to others, to our

roots, and to the future we have yet to discover. Cultivating these bonds means cultivating ourselves."

Chapter 43:

Paolo - The Value of Time

✦ ✦ ✦

Paolo sat for a long time, staring at the **eMind EEG** and the pages of his **journal**,

As if trying to **engrave this newfound clarity into his memory**.

Time had always been, for him, a **resource to be managed**,

A **sequence of hours, minutes, and seconds** to be **used efficiently**.

But tonight, with the **image of the hourglass still vivid in his mind**,

He had finally begun to **see time differently**.

He **closed his eyes**, letting his mind return to the **giant hourglass**.

The **sand flowed once again**, but this time at a **slower, more soothing rhythm**.

Each **grain of sand** seemed to **carry more than just a memory**—

It held an **emotion, a fragment of life waiting to be understood**.

Paolo stepped forward **without hesitation** and **touched another grain of sand**.

Instantly, he was transported to **a forgotten moment**—

A **spring morning**, many years ago,

Helping his **father repair an old bicycle**.

He could **smell the oil on the gears**,

See his **father's satisfied smile**

As he explained **how the chain worked**.

It was a **simple memory**, but it carried **profound meaning**—

A **moment in which time had seemed suspended**,

Free from anxiety, free from pressure.

Then, another **grain of sand** transported him to **a different scene**—

A **train slowly pulling away from the station**.

Paolo remembered standing on the **platform**, watching it **depart**.

He had known that on that train was **someone he had wanted to hold onto**.

The **feeling of loss** struck him again,

But this time, he **didn't push it away**.

He realized that **even painful moments**

Had **shaped him, molded him into the person he was today.**

Now, the **hourglass was no longer just a symbol of passing time**—

It was an **archive of his life.**

Paolo saw that **each grain, even the ones he had ignored,**

Was an **essential piece of the mosaic of his existence.**

The key was not to **control time**

Or **regret what had passed**—

But to **accept it, honor it, and value every moment.**

When he **opened his eyes**, he looked at the **graph on the screen.**

His **gamma waves continued to pulse,**

Indicating **deep cognitive activity and emotional processing.**

But he didn't need the **device to tell him that something had changed.**

He could **feel it inside himself**—

A **new awareness,**

A **perspective that pushed him to look at the future with different eyes.**

He picked up his **pen** and wrote decisively in his **journal**:

"Time is our most precious resource, not because it is limited, but because it gives us the chance to live, to choose, to grow. It doesn't matter how much time we have—what matters is how we decide to use it."

He **closed his journal** and allowed himself a **small smile**.

Tonight, Paolo had not only **reflected on time**—

He had **learned to respect it**,

To **welcome it as an ally, not an enemy**.

Chapter 44:

Francesca - The Beauty of Simplicity

✦ ✦ ✦

Francesca continued to gaze at the **sky beyond the balcony**,

The **silence around her** interrupted only by the **soft song of birds settling in for the night.**

It felt as though **nature was sending her a message**—

An **invitation to let go, to slow down**,

And to **embrace the peace she often overlooked.**

With a **gentle smile**, she put on the **eMind EEG** and closed her **eyes**,

Ready to **immerse herself in the world her mind was already beginning to create.**

Her **imagination** carried her once again to the **endless meadow**,

But this time, the **landscape seemed even more vivid.**

The **colors of the flowers glowed under a warm, golden light**,

And each **petal seemed to breathe with life**.

Francesca **walked slowly**, savoring **each step**,

Feeling the **earth beneath her bare feet**,

A **direct connection to something far greater than herself**.

When she reached the **small lake**,

The **water reflected the sky with an almost unreal clarity**.

She **knelt down**, watching the **lazy movements of the fish between the rocks**,

Following their **fluid, effortless rhythm**.

Everything around her **invited her to pause**,

To truly **see what was in front of her**.

She dipped her **fingertips into the water**,

Feeling its **coolness**,

Watching the **delicate ripples expand outward**,

As if the **lake were responding to her touch with its own quiet dance**.

Then, the **butterfly returned**.

But this time, **it was not alone**.

A **small group of pastel-colored butterflies** settled onto the **flowers nearby**,

Creating a **scene of rare beauty**.

Francesca **smiled**,

Noticing how each butterfly seemed to **move with purpose**,

Its wings **dancing to the rhythm of the wind**.

She decided to **lie down on the grass,**

Letting the **gentle sun warm her face,**

Watching the **butterflies continue to flutter above her.**

She **closed her eyes even in her imagination,**

Listening to the **sound of the wind swaying the grass,**

The **distant song of a bird,**

The **barely perceptible rustle of butterfly wings landing on petals.**

It was as though **nature was whispering an ancient secret**—

That **beauty does not need to be sought.**

It is **already there**, in **every small detail.**

When she **opened her eyes,**

The **graph on the screen** showed **predominant alpha waves,**

But also **gentle activity in the gamma waves**—

A sign that her **mind was processing a deep emotional connection**

With the **environment she had imagined.**

As she looked around,

Her **real room seemed brighter,**

As if the **world itself had taken on new life.**

She picked up her **journal** and, with **calm and fluid handwriting**, wrote:

"Simplicity is not the absence of meaning, but a pure form of beauty. It exists in small details, in silent gestures, in the things we often overlook. When we stop to truly see them, they remind us of who we are and how full of wonder the world truly is."

Francesca **closed her journal,**

A **deep sense of gratitude filling her heart,**

And remained watching the **last light of sunset,**

Knowing that **even in this quiet moment, there was all the beauty she needed.**

Chapter 45:

Paolo - The Calm After the Storm

✦ ✦ ✦

Paolo continued to watch the **raindrops sliding down the window**,

Each one tracing an **unpredictable path**.

Every **drop seemed to have its own story**,

A **unique direction**,

Yet they all eventually **merged into a single flow, finding their way downward**.

It reminded him of **his own life**—

A collection of **seemingly separate moments**,

All connected by an **invisible thread**.

He placed the **eMind EEG** on his forehead and **closed his eyes**,

Ready to **immerse himself in the inner landscape he knew was waiting for him**.

This time, it was **silence that spoke first**.

There were no **loud noises**,

Only the **whisper of the wind** and the **gentle flow of water**.

He found himself once again in the **same storm-ravaged landscape**,

But now it was **clearer, more alive**.

The **ground was soaked with rain**,

And the **air carried the scent of wet earth and renewal**.

With every **step he took**,

His **footprint briefly marked the ground**,

Only to be **filled with water moments later, erasing all trace of his passing**.

It reminded him of **how fleeting our marks on the world can be**,

Yet how **every step, no matter how small, contributes to the greater journey**.

As he walked, he noticed **a bird perched on a broken branch**.

Despite the **fragility of its perch**,

The **small creature chirped softly**,

As if **welcoming the return of calm**.

Paolo stood still, watching for a few moments,

Letting himself be **inspired by the simplicity of that scene**.

Even in **the most difficult moments**, he thought,

There is always **someone or something that finds the strength to sing**.

Moving forward, he came across **an old tree split in half.**

It **appeared as a symbol of defeat,**

But right beside its **fallen trunk,**

He noticed **new shoots pushing through the soil.**

Paolo knelt down, gently touching the **delicate green sprouts,**

Feeling a **deep connection to their quiet resilience.**

He realized that **the storm had not destroyed everything**—

It had simply **cleared the way for new growth.**

As the **sun began to pierce through the clouds,**

The **landscape around him transformed.**

Pools of water reflected a clear sky,

And the **wind carried the scent of freshness and hope.**

Paolo **stood still,**

Closed his **eyes,**

And took a **deep breath.**

This was **not just an imagined landscape**—

It was a **metaphor for his soul.**

A **terrain that had weathered the fury of the storm,**

Yet was now **ready to bloom again.**

When he **opened his eyes,**

The **graph on the screen showed a balance of gamma and alpha waves.**

A sign that his **mind had found harmony between transformation and peace**.

As he listened to **the real-world rain outside**,

He realized it no longer sounded **melancholic**.

It had become a **soothing melody**,

As if the **world itself was welcoming him into a new phase**.

Paolo picked up his **journal** and, with a **fluid and confident hand**, wrote:

"The storm does not destroy—it transforms. It is in the silence that follows that we discover who we truly are and how strong we have become. True calm is not the absence of movement, but the presence of life."

He **closed his journal with a faint smile**,

Watching as the **rain gradually gave way to a clearer sky**.

Chapter 46:

Francesca - The Courage to Begin Again

✦ ✦ ✦

Francesca remained still for a moment, wrapped in the **calm of the morning**,

Watching as the **sunrise painted the sky in shades of soft pink and golden yellow**.

She could feel **life flowing gently around her**,

Like a **river that, after rushing through wild rapids, finally reaches a peaceful plain**.

The **breath of the new day carried a promise**—

Anything was possible.

Every path could be taken.

Determined to **capture that energy**, she placed the **eMind EEG** on her forehead

And **prepared to immerse herself in her imagination**.

When she closed her **eyes**, a **dusty road appeared before her**,

Stretching into **infinity.**

She felt the **rough texture of the ground beneath her feet,**

Heard the **gentle whisper of the wind through golden fields,**

And inhaled the **earth's warmth, rising with the first light of the sun.**

She didn't know **where this road would lead,**

But every **step forward** felt **meaningful.**

It was an **act of trust,**

A **declaration of her will to move forward,**

Even in the **face of uncertainty.**

As she walked, Francesca noticed the **small details along the way**—

A **row of red poppies swaying with the wind,**

The **melodic song of a lark breaking the silence,**

The **steady rhythm of her heartbeat, synchronized with her steps.**

These **simple moments reminded her**

That **the beauty of life did not lie in perfection,**

But in the **act of living it, one step at a time.**

The **climb came unexpectedly**, bringing with it a **sense of effort.**

Her **muscles tensed**, her **breathing deepened,**

Yet inside her, a **quiet determination grew stronger.**

Each **step uphill was a choice,**

A commitment to **not stop**,

To **push beyond the limits she had once placed on herself**.

As she reached the **summit**, a thought struck her—

It was **not just the destination that mattered**,

But the **courage to take the journey**.

Then, the **landscape opened before her**, leaving her **speechless**.

The **valley below was an explosion of color and life**—

Flowers in every **shade stretched as far as the eye could see**,

Majestic trees stood **like guardians of the land**,

And a **shimmering river danced through the greenery**.

The **sunlight played with the water**,

Creating **reflections that looked like sparks of pure energy**.

And then she **realized**—

This valley was not just **a place in her mind**.

It was a **representation of the potential within her**,

Waiting to **bloom**.

Beside her, the **indistinct figure moved gently**.

Though she **couldn't see its face**,

She felt its **presence like a steady, reassuring force**.

When it whispered—

"Every beginning is an act of courage. And every step, no matter how small, brings you closer to who you are meant to be."

Francesca felt a **strength growing inside her**.

The words were **simple**,

But they carried a **powerful truth**.

She **didn't need to know exactly where the journey would lead**.

She only needed to **begin**.

When she **opened her eyes**,

She was back on her **terrace**, bathed in the **morning light**.

The **sky was brighter now**,

And the **world seemed full of promise**.

She picked up her **journal** and wrote with a **steady hand**:

"Starting over is not failure. It is a gift. It is life's way of inviting us to grow, to explore, and to become who we are meant to be. Every step, no matter how small, is an act of hope."

She **closed her journal with a smile**,

Watching the **sun rise on the horizon**.

She was **ready for the day**,

Ready for the journey,

Ready for whatever was to come.

Chapter 47:

Paolo - The Inner Map

✦ ✦ ✦

Paolo sat in **silence for a moment**, watching as the **blue light of the eMind EEG pulsed rhythmically**, almost as if it were in tune with his thoughts.

He could still **feel the presence of the library** his imagination had revealed to him,

As if a **part of him had remained there**, wandering through its **endless shelves.**

He closed his **eyes once more**, eager to **explore further what he had just discovered.**

In his **imagination**, he found himself **back in the central hall**,

Where the **large map of light stretched across the table.**

This time, he noticed **details that had escaped him before**—

Small **symbols glowing along the intertwining paths.**

Some marked **turning points,**

Others were **crossroads where he had faced difficult choices.**

There were also **places where the lines seemed broken or lost in emptiness,**

As if representing **postponed decisions or missed opportunities.**

Paolo leaned closer, his **gaze drawn to a luminous point that stood out among the rest.**

When he **brushed his fingers over it,**

He was **catapulted into a vivid memory**—

The day he had decided to **accept his first important job.**

He saw **himself, young and full of ambition**, signing the contract **with a satisfied smile.**

But along with the **excitement**, he felt the **fear he had pushed aside**—

The fear of **not being enough**, of **disappointing those who believed in him.**

The **scene faded**, and he was back at the **map.**

This time, he touched **another point, farther away.**

A **wave of emotion** surged through him as he found himself in a **difficult moment**—

The **loss of a loved one.**

The **pain was still there,**

But beside it was something **new**—

A **realization he had never fully grasped before.**

That loss had **taught him the value of relationships,**

To **never take time for granted** with the people he loved.

Each **point on the map** was a **gateway to a different part of himself,**

A **connection between past and present.**

Paolo **realized** there were **no right or wrong paths**—

Every **road, every choice, had contributed to creating the intricate, unique network that was his mind.**

Even the **broken lines,**

The ones that represented **mistakes, doubts, or unfinished stories,**

Had a **purpose.**

They were like **empty spaces waiting to be filled with new experiences.**

When he **opened his eyes,**

The **real world seemed more vivid, more full.**

The **graph on the screen displayed a combination of gamma and alpha waves,**

A sign that his **mind had reached a balance between analysis and introspection.**

Paolo removed the **device,**

And with a **rare sense of calm,**

He picked up his **journal to record what he had just experienced.**

"The map of the mind is never complete. It is a work in progress, drawn by our dreams, our fears, and our choices. Every curve, every intersection, teaches us something. And in

the end, what matters is not arriving, but having the courage to explore."

He **closed the journal**, set down his **pen**,

And allowed himself a **small smile**.

He had discovered that **even in moments of uncertainty**,

His **inner map** would always be there,

Ready to **guide him toward new destinations**.

Chapter 48:

Francesca - The Voice of the Heart

✦ ✦ ✦

Francesca sat on the **living room floor**, wrapped in the **tranquility of the evening.**

The **soft glow of the lamp cast gentle shadows on the walls,**

While the **quiet night was interrupted only by the rustling leaves outside the window, stirred by the wind.**

She had **left a small gap open**, allowing a **light breeze to enter,**

Carrying with it the **sweet scent of flowers from the garden.**

She felt **calm**, yet a **silent curiosity stirred within her.**

Tonight, she had **one clear intention**—

To listen to something she had **never truly paid attention to before**—

The voice of her heart.

For weeks, Francesca had explored **her thoughts and memories**,

Digging through the **corridors of her mind**

In search of **answers, connections, and deeper understanding**.

But she had never considered simply **stopping**

And **listening to the purest emotions within her**—

The ones that **don't pass through words**

But **vibrate like a quiet whisper, waiting to be heard**.

She placed the **eMind EEG** on her forehead, adjusting the headband carefully.

Taking a **deep breath**, she **closed her eyes** and surrendered to the **moment's stillness**.

She allowed her **imagination to take shape**,

Guided **not by thoughts**, but by **her most authentic sensations**.

In a matter of moments, she found herself in **a bright, serene space**.

It was a **vast room**, with **white walls that pulsed gently**,

As if they were **breathing in harmony with her**.

There was **no furniture**,

Except for **a small wooden chest placed at the center of the room**.

The chest was **simple**, made of **light, smooth wood**,

Without decorations, yet **incredibly inviting**.

Francesca felt **an unexplained pull toward it**,

As if it contained **something important**,

Something that **belonged to her**, but that she had **never dared to explore**.

She knelt beside it, and with **gentle hands**, she **lifted the lid**.

Inside, there were **no objects, no memories**—

Only a **warm, radiant light**,

Glowing as if it **held a life of its own**.

The **light emitted a delicate melody**,

Soft and familiar.

It was **not a song she had ever heard before**,

Yet she **recognized it immediately** as her **own**.

It was as if it **represented the truest essence of her being**,

A part of herself she had **always protected**,

But now felt **ready to rediscover**.

As she watched the **light dance**,

Tears welled up in her **eyes**.

But they were **not tears of sadness**—

They were born from **deep emotion, from a connection she had longed for without realizing it**.

It was as if the **light was speaking to her**,

Not in words, but in pure, unfiltered emotions.

Then, from the **silence**, came a **whisper**—

Soft, yet **powerful.**

"Do not be afraid to show who you are."

"Do not try to be what others want."

"Your strength is in your authenticity, in your ability to love, to dream, and to be present."

Each **word unraveled knots inside her.**

They were **not instructions,**

But a **declaration of trust and self-love.**

She now knew that these words **came from within—**

Not from an external source, not from a memory.

It was a **confirmation that everything she needed had always been there,**

Even in her **moments of doubt and uncertainty.**

She let herself be **embraced by that feeling,**

Listening to the **melody of the light,**

Watching it **sway softly, as if reassuring her.**

She stayed in that **imaginary space,**

For what felt like **an eternity,**

Yet without **rushing, without pressure.**

It was like being held in a **peaceful embrace,**

An **invitation to trust herself and her emotions.**

When she **opened her eyes,**

The **real world welcomed her back with the same gentleness.**

Her **room was still**,

But Francesca felt **different**.

As if the **weight of expectations and doubts had faded**.

She looked at the **graph on the screen**—

It showed a **perfect balance of gamma and alpha waves**,

A sign of a **profound, sincere connection between her mind and her emotions**.

She removed the **headband**, setting it beside her with a **peaceful smile**.

She picked up her **journal** and wrote:

"The voice of the heart is always there, ready to guide us. It is not loud or demanding, but if we take the time to listen, it reminds us of who we truly are. All we need is the courage to believe in it."

Chapter 49:

Paolo - The Discovery of Happiness

✦ ✦ ✦

Paolo stared at the **warm glow of the lamp on his desk**,

Watching how its light **reflected on the smooth wood.**

The past few weeks had been an **incredible journey inward**,

An **exploration that had profoundly changed how he saw the world**,

And, more importantly, **how he saw himself.**

He had confronted **fears, memories, and desires**,

But one **question still lingered**—

One he had always **avoided asking**:

What did it truly mean to be happy?

For years, he had **equated happiness with achievement**,

With **professional success, recognition, and external validation.**

But deep inside, he **knew** those answers were **incomplete.**

Tonight, he decided to **face the question head-on**—

To **discover what happiness really meant to him**.

He placed the **eMind EEG** on his forehead, adjusting the band carefully,

Then closed his **eyes**, taking a **deep breath**.

Letting the **silence of the room embrace him**,

He allowed his **mind to create**.

Immediately, he found himself in a **vast, sunlit meadow**,

Bathed in a **golden light that seemed to caress everything it touched**.

There was **nothing artificial about this place**—

The **grass swayed gently in the warm breeze**,

The **melodic chirping of birds filled the air**,

And the **sky stretched endlessly above him**,

A **perfect blue**, with a few **soft clouds drifting lazily**.

Paolo took in the **scene before him**,

Feeling an **unexpected sense of peace settle over him**.

He looked around, noticing **the smallest details**—

The **dew shimmering on blades of grass**,

The **delicate scent of wildflowers**,

And the **rhythmic sound of the wind whispering through the leaves**.

For a moment, he remained **motionless**,

Letting his **senses fully absorb the harmony around him**.

Then, without a plan or destination,

He **began to walk**.

Each **step felt freeing**.

There was **no rush**,

No goal to reach.

With every movement, he felt as though he was **shedding invisible burdens**—

Releasing **thoughts, worries, and expectations**.

The **ground beneath him was soft**,

And the **grass brushed against his ankles**,

As if **inviting him to slow down even more**.

Paolo **realized something was shifting within him**—

He wasn't **seeking answers**,

He wasn't **solving problems**.

He was simply **being present**.

And that was **enough**.

In the distance, something **caught his eye**—

A **small, solitary tree standing among the golden grass**.

Its **thin branches were heavy with shimmering fruit**,

Glowing as if **made of light**.

Paolo felt **drawn to it**, sensing that the tree **held a meaning he had yet to understand**.

When he reached it, he reached out and **picked one of the fruits**.

It was **small, smooth, and warm to the touch**.

He studied it for a moment, then took a **bite.**

The taste was **sweet**, but it was more than just **flavor**—

Each bite **filled him with a deep sense of contentment.**

It wasn't the **fruit itself that was extraordinary**—

It was the **experience of tasting it**,

Of being **fully present in the moment**,

Of allowing himself to **savor something simple, without distraction or expectation.**

Paolo sat beneath the **tree**, still holding the fruit in his hands.

He looked at the **meadow stretching endlessly before him**,

Felt the **warmth of the sun on his skin**,

And understood a **simple yet profound truth**—

Happiness was **not something complex.**

It was not a **distant goal** or a **prize to be won.**

It was here.

In the **small, unnoticed moments.**

In the ability to **experience life without carrying the past or fearing the future.**

When he **opened his eyes**,

The **real world welcomed him back with the same calmness.**

The **graph on the screen showed dominant alpha waves**,

A sign that his **mind had reached a state of deep relaxation and serenity.**

Paolo removed the **headband**, his expression **peaceful and content**.

He picked up his **journal** and wrote slowly,

Choosing each word with care,

As if he wanted to **engrave them not just on paper, but in his heart**:

"Happiness is not a place to reach, but the way we choose to walk. It is in the details, in presence, in appreciating who we are and what we have. It exists in the simplicity we often overlook—yet it gives us everything we need."

Chapter 50:

Francesca - The Beginning of Everything

✦ ✦ ✦

Francesca sat by the **open window,**

Wrapped in an **atmosphere of serenity and clarity.**

The **morning air** carried the scent of **damp grass and freshly bloomed flowers,**

A fragrance that felt like an **announcement of change.**

She gazed at the **horizon,**

Where the **sun slowly painted the sky in shades of pink and gold.**

She knew that **this day was different—**

It marked a **new direction, a new chapter.**

It was not just the **end of a journey through her mind,**

But the **beginning of a life that now truly felt like her own.**

She picked up the **eMind EEG**, placing it on her **forehead with a familiar, almost affectionate gesture.**

This device, which had once been **just a tool,**

Had become a **key that had opened doors she had never dared explore**.

She decided that **this would be her final session**—

Not to **close a cycle**,

But to **honor everything she had discovered**.

She **closed her eyes** and let her **imagination take her where it wished**.

She found herself in a **place she knew well**,

But this time, it **felt different—transformed**.

It was a **beach she had visited many times in her mental journeys**,

But now, the **sea looked vaster, brighter**,

As if it reflected **every possibility life had to offer**.

The **waves lapped gently at the shore**,

Carrying a **soothing, almost hypnotic rhythm**.

Francesca walked **slowly along the water's edge**,

Feeling the **cool sand beneath her feet**,

The **wind playing with her hair**.

Every **step was light, free**,

As if she were **leaving behind every burden, every doubt**.

As she walked, she **noticed something in the sand**—

Footprints.

They were not just **hers**,

But **intertwined with others**,

As if someone had **walked this journey alongside her all along.**

Curious, she followed the footprints,

Feeling that they would lead her **somewhere important.**

She arrived at a **large seashell resting on the shore,**

Its surface **smooth and radiant**, as if **left behind as a gift from the ocean.**

Francesca bent down to pick it up,

Noticing how the **sunlight made it shimmer with pearlescent hues.**

The shell seemed to **pulse with life,**

As if it **held a secret message inside.**

She lifted it to her **ear**, expecting to hear

The **familiar echo of the sea.**

But instead, she heard something **different, something unique.**

A **gentle melody**, soft and delicate,

That seemed to come **not from the ocean,**

But from the **heart of life itself.**

Each **note carried a part of her story.**

She saw herself in moments of **doubt,**

When the **world felt too vast**, and her **strength felt too small.**

She relived the days when she had **found the courage to rise again,**

To **face her fears and keep going.**

Then came the moments of **joy**, of **discovery**,

The moments when she had felt **exactly where she was meant to be.**

The melody **wrapped around her like an invisible embrace**,

Filling her with **understanding, acceptance, and peace.**

Francesca **sat down in the sand**,

Closing her **eyes within her imagination**,

Letting each **note wash over her like a wave.**

And then, she **understood**—

The **melody was the voice of her soul.**

A voice she had **always longed to hear**,

But had only now **learned to listen to.**

There were no more **questions.**

No more **uncertainties.**

She realized that **she didn't need to search for answers outside herself.**

Everything she needed was **already there**—

The **courage to be herself**,

The **love for who she was**,

And the **awareness that each day brought new possibilities.**

When she **opened her eyes**,

She was back in her **room**,

Wrapped in the **soft light of morning.**

The **graph on the screen showed perfect harmony**—

A balance of **gamma, alpha, and beta waves,**

As if reflecting the **inner peace she now felt.**

She removed the **device with a peaceful smile,**

Knowing that **the journey was not over**—

It was simply **changing form.**

She was **ready to face the world**

With a **new strength,**

A **new understanding**—

That **every beginning carries infinite possibilities.**

She picked up her **journal** and wrote,

With **steady, confident strokes,**

As if each word was **a seal of everything she had learned**:

"Every ending is a new beginning. Every day is a new chance to live with courage, authenticity, and love. Happiness is not elsewhere—it is here, in the meeting point between who we were and who we can become."

She set down her **pen**, closed her **journal**,

And looked out the **window.**

The **sun had fully risen,**

Illuminating the **world with golden light.**

And with it, she felt **ready to shine as well.**

Chapter 51:

The Forum of Connected Minds

✦ ✦ ✦

Paolo had stumbled upon the forum almost **by accident**,

During one of his **evening searches on eMind EEG**.

He had typed a few **keywords into a search engine**,

And the first result that appeared was **"eMind EEG Journeys"**—

An **online community** dedicated to those who used the **device**

Not just for **technical purposes**,

But to **explore their minds, emotions, and the boundaries of imagination**.

Intrigued, he **registered**, choosing the username **"PaoloMente"**,

And began **reading through the various threads**.

The forum was a **mixture of stories, questions, and discoveries**.

Some users were **neuroscience experts**,

Discussing **brain waves and cognitive dynamics**.

Others were **ordinary people**,

Sharing **personal experiences, introspective moments, and simple curiosities**.

Some described how **eMind EEG** had **improved their focus while studying**.

Others used it to **relax after a stressful day**.

And then there were people **like Paolo**,

Who saw it as a **tool to discover something deeper about themselves**.

One evening, as he scrolled through the **new posts**,

A title immediately caught his attention:

"eMind EEG Helped Me See the Light Within Me."

The author of the post signed as **"LuceInteriore"**,

A name that carried **a poetic and mysterious touch**.

Curious, Paolo clicked on the **link** and began to **read**.

LuceInteriore described a **visualization experience**

She had during a session with **eMind EEG**.

She spoke of a **serene landscape**,

A **hill with a grand tree**,

And a **warm light that seemed to emanate from her own heart**.

That **light**, she explained,

Was not just a **mental image**,

But a **tangible sensation**,

Something that had made her feel **at peace with herself**.

Her words were **precise, yet deeply emotional**,

And Paolo found himself **reading the post multiple times**,

Trying to **absorb every detail**.

Without **overthinking**, he decided to **reply**.

He wrote:

"LuceInteriore, your story is incredible. I've experienced something similar, but in a different setting. Do you think that light represents something deeper, something that was already inside you and that the device simply revealed? I'd love to hear more."

He **didn't expect an immediate response**,

But the next day, when he reopened the forum,

He found a **notification**.

LuceInteriore had written back:

"Thank you, PaoloMente. Yes, I believe eMind EEG doesn't create anything new—it only amplifies what already exists within us. It's a mirror that reflects our essence, but it takes courage to truly look at it. Would you like to share your experience with me?"

That simple **question** opened a **dialogue**

That quickly became a **daily habit**.

Paolo and **Francesca**—because soon he discovered that the person behind **"LuceInteriore"** was named Francesca—

Began exchanging **long, detailed messages**.

Paolo wrote about his **explorations into the uncharted territories of the mind**,

His **reflections on logic and imagination**,

While Francesca described her **connection with inner light**,

The **emotions that surfaced**,

And the **moments of pure serenity that the device had gifted her**.

Each exchange **brought them closer**,

Even though they had **never met in person**.

Francesca was particularly struck by **Paolo's way of analyzing his experiences**,

His constant search for **deeper meaning**.

On the other hand, Paolo found in Francesca's words

A **sensitivity and emotional depth** that fascinated him.

One day, Francesca proposed something that **Paolo hadn't expected**:

"What do you think about meeting in person? I believe sharing these experiences face-to-face could bring a new perspective. There's a small café in my city, Caffè Aurora. It's quiet, perfect for conversation. We could meet there."

Paolo hesitated.

As much as he had **enjoyed their exchanges**,

Meeting Francesca in real life posed a **new challenge**.

He had grown comfortable **expressing himself through writing**,

Protected by the **anonymity of the screen**.

But there was something about **her invitation** that pushed him to say **yes**.

Maybe it was **curiosity**.

Maybe it was the **desire to put a face and a voice to the words that had moved him so deeply**.

"Saturday afternoon?" he replied, feeling a mix of **excitement and nervousness**.

"Perfect," Francesca wrote back. **"I can't wait."**

From that moment on, they both began to **prepare for the meeting**,

Imagining what it would be like **to finally meet in person**.

For Paolo and Francesca,

The **forum was no longer just a space for exchanging ideas**—

It had become **the starting point for something new and unexpected**.

Chapter 52:

The Meeting of Minds

✦ ✦ ✦

The **Saturday afternoon had finally arrived.**

Paolo had spent the entire morning **trying to focus on other things,**

But his **mind kept returning to one place—**

Caffè Aurora and his **first meeting with Francesca.**

He didn't know **what to expect.**

Until now, Francesca had been **a series of words on a screen,**

An **image constructed through their exchanges.**

But he had no idea **who he would truly meet in person.**

He dressed **simply but thoughtfully,**

Choosing a **light blue shirt and dark jeans.**

He wanted to **feel comfortable,**

Yet also convey a **sense of seriousness.**

Taking the **train to Francesca's city,**

He tried to **ignore the nervous excitement**

That grew with **each passing kilometer**.

The First Encounter

Caffè Aurora was **exactly as Francesca had described**—

A **small, cozy café**,

With **pastel-colored walls and large windows**

That let in the **soft afternoon light**.

Inside, there were **only a few tables**,

Each decorated with **white tablecloths**

And **small vases filled with fresh flowers**.

Paolo glanced around, **looking for Francesca among the seated guests**.

He didn't have to **wait long**.

A **figure stepped through the doorway**,

Her smile **shy but sincere**.

It was **Francesca**.

She wore a **light dress**,

A color that reminded him of **the sky just before sunset**.

A **shoulder bag hung at her side**,

And her **eyes sparkled with a warm, gentle light**.

Paolo knew **instantly that it was her**.

"Paolo?" she asked, stopping in front of him.

"Yes," he replied, standing and extending his hand.

"Francesca?"

She nodded, shaking his hand.

"It's so nice to finally meet you in person."

"For me too," Paolo said with a small smile.

They sat at a **table by the window**.

For a moment, a **comfortable silence settled between them**.

Not awkward, but the kind that happens **when two people are still adjusting to each other's presence**.

Breaking the Ice

"Isn't it strange?" Francesca said, finally breaking the silence.

"We've talked so much online, and yet, here we are, like two strangers meeting for the first time."

"I know," Paolo chuckled softly.

"But in a way, I don't feel like a stranger. It's like I already know a part of you."

Francesca smiled, **tracing the edge of her tea cup with her fingers**.

"I've felt the same way. It's strange, but at the same time, it feels familiar."

They started talking **naturally**,

As if the **forum had only been a prelude** to this conversation.

Paolo spoke about **how he had discovered eMind EEG**

And how it had **changed his perspective on life**.

Francesca, in turn, shared her **own experiences**,

The sessions where she had **found her inner light**,

The moments when the **device had helped her make peace with herself**.

The Power of Connection

"There was a moment," Francesca said, resting her chin on her hand,

"when I realized that eMind EEG wasn't just a technological tool. It was during a session when I felt completely connected to a part of me I had never listened to before. Like my mind and my heart were finally speaking the same language."

Paolo nodded, **understanding exactly what she meant.**

"It was the same for me. It's like every session helps me discover another piece of the puzzle. I haven't completed it yet, but now I know I don't have to be afraid to look inside."

They talked for **hours**,

Losing track of **time**.

They **compared their experiences**,

Finding **surprising similarities** in their **inner journeys**,

But also **differences that made their conversation even richer**.

Paolo was more **rational**,

Always **analyzing every detail**,

While **Francesca followed intuition and emotions**.

Yet somehow, their **different approaches complemented each other perfectly**.

A New Beginning

As the **sun began to set**,

Filling the **café with a golden glow**,

Francesca looked at Paolo and smiled.

"You know," she said, **"I never thought a simple forum could bring me here, sitting across from someone who truly understands what all this means."**

"Neither did I," Paolo admitted, leaning back in his chair.

"But maybe it was meant to happen. Maybe we were supposed to meet and share this journey."

Francesca's smile grew.

"Maybe we were."

They left the café together,

Walking **slowly through the quiet streets of the city.**

They didn't know **what the future had in store**,

But one thing was certain—

This day **marked the beginning of something special.**

A **connection born from the desire to explore the depths of the mind and the heart.**

Chapter 53:

Paolo's Proposal

✦ ✦ ✦

After their **first meeting at the café**, Paolo and Francesca continued **exchanging messages**

And sharing **insights from their eMind EEG experiments.**

Every conversation seemed to **open new doors**,

Deepening their **shared curiosity.**

One evening, as they discussed ideas on the **forum**,

Paolo typed a message that revealed his **latest realization**:

"Francesca, I've been thinking... why don't we try an experiment together? I believe working closely might help us discover something we can't achieve alone. If you're interested, you could come to my place. I have the space and all the equipment to explore this further."

Francesca read the message on her **computer screen**,

A **subtle smile curving her lips.**

The idea **intrigued her.**

Working with Paolo seemed like the **perfect opportunity**

To **not only deepen their understanding of eMind EEG**

But also to explore the **special connection she felt between them**.

She replied:

"That's an interesting idea. It would be great to combine our insights. When do you want to do these experiments?"

A few minutes later, Paolo's response appeared:

"This Saturday afternoon, if you're free. I can prepare a series of tests and see how our brain waves react in shared scenarios. But I'll warn you—I'm also very curious about your visualization techniques. Maybe you can teach me something."

Francesca laughed softly to herself.

Paolo was **direct**,

But it was clear that his **proposal wasn't just technical**—

There was a **genuine desire to learn more**,

Not just about the **device**,

But about **her**.

She typed back:

"Saturday sounds perfect. I'll prepare a few ideas for us to try together."

Chapter 54:

The Meeting at Paolo's Home

✦ ✦ ✦

On the agreed-upon day, **Francesca arrived on time**,

Carrying a **small bag** with her **eMind EEG device, a notebook filled with notes, and a few written ideas for their experiments.**

Paolo greeted her at the door with a **warm smile**,

Wearing a **gray sweater and a pair of comfortable jeans.**

His home felt **welcoming**,

Bathed in **natural light streaming through the large windows of his study**,

Where he had set up **everything needed for their experiments**:

Two **chairs**,

A **table with their laptops**,

And an entire **wall covered with charts and notes.**

"Welcome to my improvised lab," Paolo said, gesturing toward the room.

"It may not be high-tech, but it has everything we need."

Francesca smiled, setting her **bag down beside the chair**.

"It's perfect. I think we're going to make some interesting discoveries together."

Preparing the Experiment

As they prepared the **devices and equipment**,

They started talking about **their personal experiences with eMind EEG**.

Paolo showed Francesca the **charts from his recent sessions**,

Explaining how his **brain waves changed during deep concentration and meditation.**

Francesca, in turn, shared her **visualization techniques**,

And how using the **device had helped her find clarity and emotional relief.**

Once everything was **set up**, Paolo suggested their **first experiment**:

Recording their brain waves while imagining the same scene.

"I want to see if our minds can synchronize, at least in part," he said,

His voice filled with **enthusiasm and curiosity**.

They sat **side by side**, their **devices connected to their laptops**.

Paolo suggested an **image for them to visualize together**—

A **beach at sunset**,

With the **sound of waves crashing against the shore**

And the **warmth of the sun on their skin**.

They **closed their eyes**,

Letting the **scene unfold in their minds**.

A Shared Vision

The **room fell silent**,

The only sound was the **faint hum of their computers**.

Francesca felt something **unusual**—

A strange **connection**,

As if the **beach wasn't just in her imagination**,

But a **real place they were sharing together**.

She could **almost feel Paolo's presence there**,

Not in the **physical world**,

But within the **space of their collective consciousness**.

When they **opened their eyes**,

They turned toward the **screen to analyze the data**.

Their **expressions shifted from curiosity to amazement**.

Their **brain wave patterns had synchronized** at multiple moments.

"I don't believe it," Francesca murmured, staring at the screen.

"Our gamma and alpha waves aligned several times. It's like our minds were really on the same beach."

Paolo nodded, his eyes shining with excitement.

"This is incredible. I never thought it was possible to reach such a deep level of connection with someone."

Chapter 55:

A Discovery Beyond Science

✦ ✦ ✦

The **experiment continued**,

But Paolo and Francesca soon realized that the **real discovery**

Was **not just in the data or the graphs**.

It was in the **connection forming between them**,

A **synchronization that went beyond eMind EEG**.

That day was not just **a step forward in their research**—

It was the **beginning of something much deeper**.

Francesca smiled as she watched Paolo **organizing the devices**.

"You know, I think the real discovery isn't in the brain waves. It's in the connections we create—with others and with ourselves."

Paolo looked at her, struck by her words.

"You're right. And maybe this is what we've been searching for all along—without even knowing it."

An Unfinished Conversation

That evening, as **Francesca prepared to leave**,

Both of them knew that **this wasn't just about experiments anymore**.

It was the **start of a new adventure**,

A journey that would lead them to **discover not only the potential of eMind EEG**

But also the **power of human connection**.

Paolo observed Francesca as she **adjusted her scarf before stepping outside**.

It was a **simple moment**,

Yet **it held a deep significance**.

After everything they had **experienced and explored together**,

Their **bond** felt stronger than ever.

Francesca paused for a moment **at the door**,

Turning back to him with a **thoughtful expression**.

"You know, I've been thinking a lot about what you told me the other day," she said.

"About how the mind can create real experiences through imagination."

Paolo nodded, a **small smile playing on his lips**.

"It's something I've studied for a while," he admitted.

"I've gathered data, tracked my brain waves while imagining sensations and emotions."

She looked at him, intrigued.

"You kept records of everything?"

Paolo picked up a **tablet from the table and powered it on.**

"Yes. I wanted to understand if imagining a connection, a feeling, could have a measurable effect on the brain. And you know what? The results are surprising."

Francesca stepped closer, resting a hand on his arm.

"Then why not share this? It could help so many people understand the power of their own minds."

Paolo inhaled deeply, considering her words.

"You're right. If it can help someone unlock a part of themselves, then it's worth sharing."

Francesca smiled.

"So why not start now? Tell me everything from the beginning."

A Deeper Understanding

Paolo looked at her for a moment, then nodded with determination.

"Alright. I'll walk you through my research step by step. Here's how it all started□"

Francesca sat down beside him, curiosity **lighting up her eyes**.

Paolo began scrolling through the files on his **tablet**,

Displaying the data he had collected over **weeks of self-exploration** with **eMind EEG**.

"Look at this," he said, pointing to a graph on the screen.

"This is my brain activity during a guided visualization. See the difference between my resting state and the moment I imagined an emotionally intense scene?"

Francesca studied the graph carefully.

"That's incredible□ It's as if the brain truly reacts to something that isn't physically happening."

Paolo nodded.

"Exactly. That's the core of my study. Imagination isn't just a mental exercise—it has measurable effects on our mental state and emotions."

They continued discussing the **implications of his research,**

Analyzing **patterns and possibilities,**

Both fascinated by what they were uncovering.

But then, Francesca **yawned**, realizing how **late it had gotten.**

"Paolo, this is all incredibly fascinating, but I really need to go. Promise me you'll keep me updated on your progress."

Paolo smiled.

"I promise."

Francesca stood up, adjusting her **scarf once more,**

And headed toward the **door**.

"See you soon. Keep exploring this—I want to hear everything."

"You can count on it," Paolo replied.

The **door closed softly behind her.**

Paolo's Realization

For a few moments, Paolo **stood still,**

His **mind racing with thoughts.**

The **conversation with Francesca** had given him **the final push he needed.**

This was no longer **just an experiment.**

It was no longer **just personal curiosity.**

He had something **real in front of him**—

Something that could **reshape how people understood their own minds.**

He walked over to his **desk**, sat down, and opened his **tablet.**

The data from the past few weeks was **all there, waiting to be analyzed.**

This was no longer **just a theory.**

Imagination was not **merely a trick of the mind**—

It was a **powerful tool capable of shaping reality itself.**

With a **deep breath,**

He **began to write.**

Imagining Love:

Scientific Analysis and Personal Experimentation

(Appendix to the novel "Imagine Being Connected")

Connection between Narration and Study

The cursor was flashing on the tablet screen. Paolo took a moment to think, then he started typing.

"To better understand this phenomenon, I structured detailed research that combines neuroscience and personal experience. What you will find below is the method I followed, the data collected and the possible implications of this study."

What follows is the result of my study: an analysis of how imagination can influence our mind and our emotions. If you are curious to explore this possibility, continue reading

1. Introduction

To conduct this study, **eMind EEG** was used, an advanced device for monitoring brain waves. This tool allows to detect brain activity in real time through EEG sensors, providing accurate data on the different frequency bands, in particular **Alpha and Theta**, essential to understand the effects of imagination on emotions and mental state.

The data analysis was performed using **eMind APP**, a specialized software that allows you to view, record and export EEG data for in-depth analysis. Thanks to this application, it was possible to track and compare the variations in brain waves in different mental states, observing how the imagination of love influences brain activity.

This combination of EEG technology and analysis software made it possible to study the link between imagination and brain activity in a more in-depth and measurable way.

Why study the imagination of love?

Love is a powerful emotion, capable of profoundly influencing our mental and physical state. But what happens in the brain when we imagine being in love?

This study was born from the desire to explore the link between imagination and emotions, using technological tools to measure their effects on brain waves. The goal is to understand whether the simple act of imagining feeling love can generate a measurable response in the brain, influencing mood and general well-being.

Although love is often considered a subjective experience, it may have concrete neurophysiological correlations.

This research aims to analyze how the brain reacts when a person imagines being in love, collecting data through an EEG device to identify any recurring brain patterns.

This research is presented as an appendix to the novel *Imagine Being Connected*. Although the novel does not deal directly with imagined love, its themes of connection and mental interaction fit well with the idea of exploring the power of the mind and imagination on brain activity.

Objective of the Manual

This manual guides the reader in exploring the connection between imagination, mind and technology, combining subjective experience and scientific measurement. The main objectives are:

To analyze how the imagination of love affects brain waves and mental state.

To create a bridge between subjective perception and neuroscientific data, through EEG measurement tools.

To provide a practical method to experience first-hand the effect of the imagination of love on one's mind.

To suggest concrete applications in personal improvement, meditation and emotional well-being.

2. Scientific Foundations: Brain Waves and Emotions

- Brain Waves and Their Meaning

Brain waves are electrical oscillations generated by neuronal activity. They are divided into five main categories:

Delta (0.5 - 4 Hz) → Deep sleep and regeneration.

Theta (4 - 8 Hz) → Creativity, dreams and meditative states.

Alpha (8 - 14 Hz) → Relaxation and mental calm.

Beta (14 - 30 Hz) → Concentration and intense cognitive activity.

Gamma (30 - 100 Hz) → Complex cognitive processes and awareness.

- Imagination and Brain Response

Imagination is not just a psychological phenomenon, but a concrete activity that activates specific brain areas. Previous studies have shown that imagining a lived scene can generate neural responses similar to those of a real experience.

This observation leads to the central question of our study: can imagining love induce specific brain responses that reflect a state of well-being and emotional involvement?

3. The EEG Device and the Technology Used

Device Choice: Why **eMind EEG**?

eMind EEG is a portable EEG device that allows you to monitor brain waves in real time. Compared to other consumer EEGs, **eMind EEG** offers an optimal balance between accuracy, accessibility and ease of use.

eMind APP: a powerful analysis tool

The eMind APP application allows you to access raw EEG data and analyze them in a in-depth. Its main features include:

Real-time visualization of brain waves.

Recording and exporting EEG data for later analysis.

Application of custom filters to highlight specific frequency bands.

Integration with third-party software for advanced data analysis.

4. Study Protocol: How to Test the Imagination of Love

- Detailed Methodology

Baseline Phase (5 minutes) → Record the neutral brain state without thinking about anything specific.

Imagination Phase (10 minutes) → Mentally construct a romantic scene and focus on the emotions experienced.

Recovery Phase (5 minutes) → Observe how the brain gradually returns to the neutral state.

Repeat the test for multiple sessions → Verify the consistency and repeatability of the results.

5. Results and Data Interpretation

- Advanced Analysis

Creation of comparison graphs between baseline, imagination and recovery phases.

Predominant brainwaves assessed based on intensity of imagery.

Comparison between different subjects to identify common patterns.

Possible Applications

Love-focused meditation to improve emotional well-being.

Use of visualization techniques to stimulate creativity and relaxation.

Therapeutic use for managing stress and loneliness.

6. Conclusions and Suggestions for Practical Application

The experiment shows that imagery of love significantly affects brain activity.

A systematic practice can be developed to improve emotional well-being through positive visualization.

Practical suggestions for applying these findings in daily life:

- Create a guided imagery routine for relaxation.
- Use visualization in combination with controlled breathing.

- Integrate the use of EEG devices to monitor your progress.

7. References and Sources

Here are some research and scientific studies that support the topics covered in this manual:

Brainwave Studies and Imagery

- **Davidson, R. J., & McEwen, B. S. (2012).** "Social influences on neuroplasticity: Stress and interventions to promote well-being." *Nature Neuroscience, 15(5)*, 689–695.
- **Kosslyn, S. M., Ganis, G., & Thompson, W. L. (2001).** "Neural foundations of imagery." *Nature Reviews Neuroscience, 2(9)*, 635–642.
- **Amodio, D. M. (2019).** "Social cognition 2.0: An interactive memory systems account." *Trends in Cognitive Sciences, 23(1)*, 21–33.

EEG and neurofeedback in mental visualization

- **Cohen, M.X. (2014).** "Analyzing neural time series data: Theory and practice." MIT Press.
- **Miller, R. (2007).** "Neurofeedback and self-regulation in cognitive enhancement." Journal of Neuroscience Research, 85(4), 785–793.
- **Lutz, A., Slagter, H. A., Dunne, J. D., & Davidson, R. J. (2008).** "Attention regulation and monitoring in meditation." Trends in Cognitive Sciences, 12(4), 163–169.

Emotional imagination and brain states

- **Singer, T., Seymour, B., O'Doherty, J., Kaube, H., Dolan, R. J., & Frith, C. D. (2004).** "Empathy for pain involves the affective but not sensory components of pain." *Science, 303(5661),* 1157–1162.

- **Schneider, S., Brummer, V., Abel, T., Askew, C. D., & Strüder, H. K. (2009).** "Changes in brain cortical activity measured by EEG are related to exercise-induced blood lactate levels." *Neuroscience Letters, 452(2),* 121–125.

Acknowledgements and The Role of Artificial Intelligence

This manual was developed with the support of advanced artificial intelligence, which contributed to the structuring of the protocol, the analysis of EEG data and the organization of scientific information. The use of AI allowed to optimize the communication of results, improve methodological coherence and make the entire study more accessible and understandable. Its contribution helped to transform an idea into a structured and scientifically valid research method.

The integration of **technology, neuroscience and imagination** opens up new perspectives of study, offering increasingly advanced tools to understand the link between brain activity and subjective experience. This manual is just a starting point: future research will be able to further deepen the role of AI in EEG analysis, exploring new application scenarios in the field of cognitive psychology and guided meditation.

Analysis of EEG Waves in the Three Phases

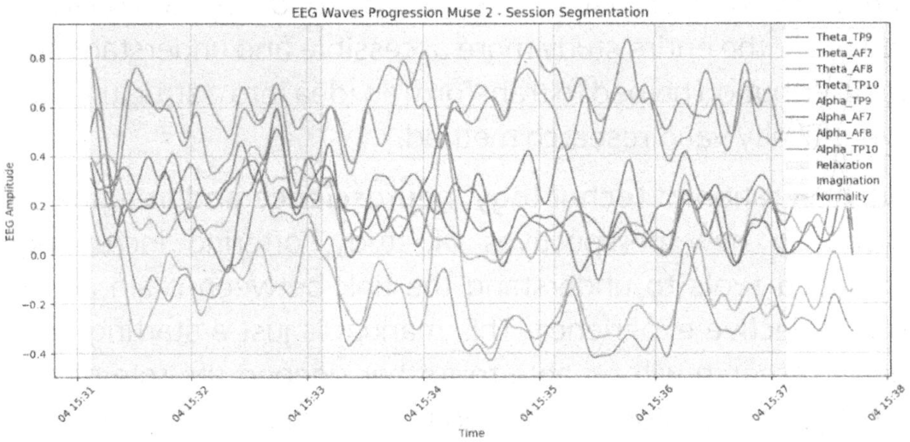

We observe how brain waves behave in the three different phases of the test:

1. Relaxation (Blue)

Duration: First 2 minutes

Main features:

- **Alpha waves predominant** (especially **Alpha_TP9** and **Alpha_TP10**): indicate a state of calm and relaxation.
- **Theta waves present but contained**: sign of a relaxed mind, with slight spontaneous imaginative activity.
- **Few sudden peaks**, suggesting stable brain activity.

📌 **Interpretation:**

The high presence of Alpha waves confirms that the brain was in a relaxed state, without strong external or internal stimuli. This is typical of light meditation or a state of mental rest.

🎭 **2. Imagination (Green)**

Duration: Minutes 2-4

Main features:

- **Significant increase in Theta waves** (especially **Theta_TP9** and **Theta_TP10**), indicating intense imaginative and emotional activity.
- **Alphas still present, but more unstable**, signalling greater cognitive involvement.
- **More pronounced peaks and swings**, especially between Theta and Alpha, suggesting changes in mental state.

📌 **Interpretation:**

The increase in Theta waves is consistent with a state of deep imagination. This is a positive sign confirming the effect of imagination on brain activity.

🔄 **3. Normality (Red)**

Duration: Minutes 4-6

Main features:

- **Reduction of Theta waves**, signal that imaginative activity is decreasing.
- **Alpha waves return to levels similar to the relaxation phase**, signalling a return to mental stability.

- **More contained oscillations**, with fewer sudden peaks.

📌 **Interpretation:**

The brain is returning to a more neutral state after imaginal activation. The slowing down of Theta waves and the restoration of Alpha waves suggest that the mind is re-establishing a balance.

📌 **General conclusion**

☑ **Theta waves dominated during imagination, confirming the effect of emotional mental activity.**

☑ **The relaxation phase had a high presence of Alpha, as expected for a state of mental calmness.**

☑ **The return to normal showed a gradual recovery, without excessive oscillations.**

Made in the USA
Monee, IL
04 November 2026

67377417R10148